Ocean Space Rights

Lawrence Juda

foreword by
Oliver J. Lissitzyn

The Praeger Special Studies program—utilizing the most modern and efficient book production techniques and a selective worldwide distribution network—makes available to the academic, government, and business communities significant, timely research in U.S. and international economic, social, and political development.

Ocean Space Rights
Developing U.S. Policy

PRAEGER SPECIAL STUDIES IN INTERNATIONAL POLITICS AND GOVERNMENT

Praeger Publishers New York Washington London

Library of Congress Cataloging in Publication Data

Juda, Lawrence.
Ocean space rights.

(Praeger special studies in international politics and government)
Based on the author's thesis, Columbia University.
Bibliography: p.
Includes index.
1. Continental shelf. 2. Continental shelf—United States. I. Title.
JX 4143.J83 341.44'8 74-1732
ISBN 0-275-09240-2

PRAEGER PUBLISHERS
111 Fourth Avenue, New York, N.Y. 10003, U.S.A.
5, Cromwell Place, London SW7 2JL, England

Published in the United States of America in 1975
by Praeger Publishers, Inc.

Printed in the United States of America

FOREWORD

by Oliver J. Lissitzyn

Since World War II, international law has been in a state of particularly rapid flux. Accelerating technological progress and profound political changes in the world community, of which the emergence of the Third World is only the most obvious manifestation, have put in question or are rendering obsolete many of the old and once firmly established norms and have at the same time created new needs, demands, and expectations addressed to international law.

Like all legal systems, the law of nations is a product of the political processes within the community whose needs it serves for accommodation of diverse interests and for regulation. Law is formalized policy. International law is primarily a crystallization of a rough consensus among nations—a consensus arrived at in a process of contention, compromise, and agreement of the entities known as sovereign states. Each state seeks to defend and advance its own interests and policies in this complex process of interaction. States, of course, are unequal in many political, economic, and military respects, and this inequality is reflected, sometimes very subtly, in the weight of the input of each of them in the international law-making process.

The relation of this process to international politics and to the distinctive interests and policies of the various states has come to be widely understood. But this process and particular aspects of it have been generally studied without much, if any, reference to the ways in which each state's own interests and policies, in their relation to international law, come to be shaped. Nation-states are still too often viewed as atom-like entities, with little or no attention paid to the political forces and processes within each of them. Yet a state's policy on the international arena is rarely, if ever, an expression of a monolithic and dissension-free domestic consensus. In most if not all states, foreign policy, including the policy with respect to the formation and application of international law, is itself a product of contention and accommodation of a number of different interests and views. The domestic process by which the state's policy or position on a problem of international law is formulated may be quite complex. It may be the result of pressures generated by diverse governmental and nongovernmental interest groups whose power to influence the official formulation of policy may fluctuate during the different stages of the process and may be exerted by a variety of means and in several loci, for example, in the executive as well as in the legislative branches of government. The way in which particular norms of international law are shaped and come into being, often superseding older rules, cannot be completely understood without taking into account the domestic political processes that shape the policies officially espoused by the governments of all the nation-states participating in the world political process. Until now, however, international law scholars, including political scientists with a special interest in the law of nations, have generally failed to engage in serious research studies of this substratum of the international law-making process.

It is, of course, evident that no one scholar or group of scholars can conduct such studies throughout the full spectrum of international law and with respect to all the nations of the world. But it is possible to make a start by a case study of the policy-making process within a single nation-state in relation to a particular international law topic of contemporary significance. This is what Professor Juda has done, with enlightening results, in this volume. He has selected for his study the development of the American policy with respect to the continental shelf. This choice is justified by several considerations.

First, while the need for legal regulation of man's activities in such entirely novel environments as outer space has been obvious and has already led to the conclusion of several important agreements, in terms of immediate practical significance and the multiplicity of the interests involved the law of outer space is overshadowed by the often conflicting claims and expectations addressed to the law regulating activities in an environment long familiar to man—the sea. Long a highway for navigation, commerce, and war, as well as a source of food and some other products wanted by man, the sea has acquired in the twentieth century additional importance as technology has made increasingly feasible the exploitation of vast mineral resources on and under the ocean floor. Burgeoning populations and improved fishing techniques continue to enhance the importance of the living resources of the sea. The development of new weapons and weapon delivery systems, such as nuclear-armed submarines, and of more sophisticated submarine detection devices, has complicated the problems of the uses of the seas and their beds for military purposes. The rapid expansion of industry in many countries and the growth of population have resulted in new problems of control of pollution as the seas have acquired greater significance as dumping places for various wastes and as the frequency of spills of oil and other noxious substances has increased. The proliferation of the uses of the sea has in turn resulted in greatly enhanced scientific interest in the exploration of the marine environment in all its aspects, leading to restrictions on such exploration by some coastal nations fearful of foreigners deriving undue economic or military benefits from it.

The international community, recognizing the need for new or modified legal regulation of the uses of the sea, has resorted to a variety of means of developing the law, including notably the 1958 United Nations Conference on the Law of the Sea at Geneva, which produced four conventions. One of these, on the continental shelf, was designed to regulate the exploration and exploitation of the natural resources of the seabed and subsoil of submarine areas adjacent to the coasts of states by recognizing that the latter had sovereign rights of control and jurisdiction over such activities. But this convention, never ratified by a majority of states, gave rise to new uncertainties and controversies, and its inadequacy became manifest when technology was about to make possible the exploitation of mineral resources of the ocean bed at great depths and at great distances from land.

The recognition of the resulting need for new regulatory measures, as well as new claims and policies of many states with respect to other aspects of the marine environment, led the General Assembly of the United Nations to call another international conference on the law of the sea which, after a brief initial meeting in New York late in 1973, began its substantive work in the summer of 1974 in Caracas, Venezuela. Although the agenda of the conference covered a large number of the problems of the law of the sea, of crucial importance was the effort to reach an agreement that would recast or supplant the Continental Shelf Convention of 1958. An understanding of the processes by which the legal doctrine of the continental shelf developed is, therefore, not only of importance as throwing additional light on the general international lawmaking process but may be also expected to illuminate the ongoing efforts to modify or discard this doctrine by providing for more adequate regulation of the exploration and exploitation of the riches of the ocean bed.

The second justification for Professor Juda's choice of topic is the initiative of the United States in the formulation of continental shelf doctrine. American policy, initially strongly linked with the desire of some interest groups to protect from foreign exploitation certain stocks of fish in waters beyond the three-mile territorial waters of the United States, crystallized in President Truman's proclamation of September 1945 claiming for the United States the exclusive control and jurisdiction over the natural resources of the "continental shelf" contiguous to the United States. This proclamation, although stating a claim of dubious standing in then existing international law, met with no resistance in the international community and set the precedent for similar and even more far-reaching claims by a steadily increasing number of other states, culminating in the recognition and attempted formulation of the new doctrine in the 1958 convention. However, in the meantime, and also after the 1958 conference, many problems of interpretation and application of the doctrine kept arising. The policy of the United States eventually had to be reconsidered when the international community began to move toward a new general conference on the law of the sea. In the international arena, the influence of the United States in the process of reaching a new consensus is still of utmost importance. Without the consent or acquiescence of the United States, no new international regime for the seabed is likely to be created, or, if created, to be viable.

And third, the United States is uniquely suitable for a case study of this kind not only because of its international influence, but also because it offers unequalled opportunities for a private scholar to obtain large masses of official and unofficial source materials on the domestic policy-making process closely related to the nation's foreign relations. In no other country could he obtain access to comparably extensive materials. Not only in closed societies ruled by authoritarian regimes, but even in most Western liberal democracies, governmental agencies and officials, as well as private interest groups, are much more secretive with respect to policy-making processes, particularly as they relate to foreign affairs, than they are in the United States.

Professor Juda, who is well versed in both international law and American politics, has accomplished his task with superb resourcefulness and thoroughness. This is not the place to summarize the very complex picture of the decision-making process he presents. His work is unique in its careful and detailed delineation of the tortuous processes and of the many contributing influences by which the American continental shelf policy, from its very origin, was formulated, subsequently modified, and eventually reconsidered.

This study deserves the close attention not only of those who are following the contemporary developments in the law of the sea, but of all those, lawyers and political scientists alike, who are seriously interested in fully understanding international lawmaking, including its roots in the domestic political processes.

ACKNOWLEDGMENTS

I should like to express my deep appreciation to a number of people and organizations without whose cooperation this work could not have been completed. To the many present and former government officials and representatives of private interests who granted me interviews or corresponded with me, I give my sincere thanks. Their assistance was invaluable. I am grateful for the comments and suggestions of Professor Louis Henkin of Columbia University on portions of an early draft of this book. I am particularly indebted to Professor Oliver J. Lissitzyn of Columbia University whose enthusiasm for scholarly research served as a major source of inspiration and stimulus and who generously shared with me his valuable time and his vast knowledge of the international lawmaking process. I am also appreciative for the financial assistance provided by Muhlenberg College for the preparation of this manuscript. Finally, and very important, my sincere thanks to Alice and Emily.

CONTENTS

Ocean Space Rights

CHAPTER

1

INTRODUCTION

Students of international relations have long observed that advances in technology profoundly affect the means whereby states interact, in terms of both peaceful and warlike activities. Advances in areas such as mass production, transportation, communications, and weaponry have clearly left their mark on international relations in the past centuries. While technological change affects the means of state interaction, it can also introduce new areas of potential conflict or cooperation. In the second half of the twentieth century two vast new worlds are opening up to man because of technological advances. Exploration of outer space is no longer a subject merely for man's imagination. At the same time, and of greater immediate and practical importance, exploration of the "inner space" of the vast oceans of the world and the exploitation of the resources found in the ocean, the seabed, or in the subsoil beneath the oceans is also becoming a reality.

The continents of the world are in part situated above water level. Gently, or not so gently, the far edges of the continents slide beneath the water that covers most of the earth's surface. Part of this extension of the land mass under the sea is the continental shelf. The continental shelf may be described as the submarine continuation of the land mass extending under the sea to where the sea floor begins to fall off more steeply into oceanic depths. The original concept of the continental shelf is of geological, geographic, and oceanographic origin.[1] It was only with the 1945 Truman Proclamation that the continental shelf came into being as a legal concept. At the time, the proclamation provided a departure from then existing international law, which had not regarded the seabed outside of territorial waters as appertaining to the coastal state.[2]

With advances in the technology of the fishing and petroleum industries the continental shelf assumed an importance that required the attention of national decision makers. In addition to the long running dispute between the United States and a number of Latin American states on the scope of the continental shelf and on rights in regard to the epicontinental waters, disagreements between other states have erupted in connection with the continental shelf. Conflicts arising from claims to offshore areas have involved a number of states including Denmark, the Netherlands, and West Germany; China and Japan; Japan and Australia; the Soviet Union and Japan; South Korea and Japan; France and Brazil; and Greece and Turkey. In the face of recent developments and concern expressed about offshore areas, the United Nations General Assembly decided to convene a diplomatic conference on the law of the sea to begin in 1973.[3]

Today the right of coastal state jurisdiction over the continental shelf is universally recognized, though with some important variations as to what is entailed by such jurisdiction. As of 1967, the shelf areas off the coasts of the United States accounted for approximately 12 percent of America's oil production and 10 percent of its natural gas production.[4] More recent figures suggest even greater importance; in terms of world wide production it has been estimated that about 18 percent of the total supply of petroleum comes from offshore resources. It has further been predicted that by the year 2000 as much as 40-50 percent of the world's petroleum production will be derived from offshore sources.[5]

As important as the continental shelf is for its oil deposits and its actual and potential military uses, the matter of the continental shelf has further important ramifications. While certain states view the continental shelf primarily in terms of mineral rights and mineral exploitation, other states see it primarily in terms of the exploitation and conservation of the living resources of the epicontinental sea. Some states go so far as to claim sovereignty over the waters above the continental shelf. This raises a fundamental question: for what resources and for what purposes does the legal concept of the continental shelf exist in international law? The answer to this question has far reaching implications for the legal regime of the seas, for the economies of a number of states, for the wise and efficient use of the ocean's varied resources, and for the military posture of some states.

The United States has long been a champion of the doctrine of freedom of the seas, insisting that to the greatest extent possible the oceans be kept free from exclusive national claims of jurisdiction. It has been a long held feeling that the greater the freedom of the seas, the easier it will be for the United States to make its power felt. In addition, some American fishermen also desire maximum freedom of the seas so they can fish wherever the fish are. They would prefer to be restricted in their search for fish only by narrow bands of nationally controlled waters adjacent to foreign states. Aware of this, the drafters of the Truman Proclamation on the Continental Shelf were careful to deal only with the mineral resources of the shelf.

This limited American claim, even if reciprocated by all other states, would allow the United States to benefit from the rich oil-bearing continental shelf off its coasts, yet would provide no bar to American fishermen operating close to the coast of other states. But the Truman Proclamation raised an important question: how can one sovereign state have the right to unilaterally claim exclusive jurisdiction over minerals that lie off its coasts and yet tell other sovereign states that they cannot make similar claims to nonmineral resources that may be located an equal distance off their coasts? An American claim to the mineral rights on a wide continental shelf area, however defined, may lead to wide claims by other states to secure for themselves whatever resources they believe exist off their coasts, be they mineral or otherwise.

In terms of the development of the continental shelf concept, it has been observed that the very process of extracting minerals or petroleum must to some degree effect fishing and navigation rights. Will not the coastal state claim some right of jurisdiction for security purposes in the area where exploitation is taking place? Will not the coastal state claim the right to take those actions necessary to protect the ecology of the work area and its surroundings? Will not steps be taken by the coastal state in the name of navigational safety? In general, the question must be asked: do we not, by national claims to the continental shelf for mineral exploitation rights, unleash forces which will steadily erode the doctrine of freedom of the epicontinental seas? It has been suggested by some observers that jurisdictional claims for one purpose in ocean areas eventually lead to claims for other purposes.[6] Has this consideration, referred to as "creeping jurisdiction," been taken into account by those who determine American actions with respect to the continental shelf? It may be asserted that the United States no longer needs to support the concept of the freedom of the seas in light of present assessments of national interests.* The question is, however, whether the expansion of claims

---

*"National interest" as used here refers to the resultant of the continuing political process of conflict and compromise through which positions of the United States government are determined. This "national interest" is subject to change as perceptions of actors in the political process change, as the available information changes, as the relative strengths of various actors changes, as new actors with varying viewpoints enter the policy-making process, and as developing technology opens up the possibility of new alternatives.

to continental shelf areas is seen and considered in the light of possible threat to the freedom of the seas.

Another problem arising from the development of the continental shelf doctrine is that of the definition of the shelf in terms of its horizontal extent. Here we are not asking for what purposes a state may be said to have jurisdiction and control over the continental shelf, but rather how far may this area extend from the coast of a state. Is the legal continental shelf merely the geographical shelf? Does it encompass the entire continental margin?* Can it extend, in accordance with a very liberal interpretation of the exploitability clause of Article 1 of the 1958 Continental Shelf Convention, to the midpoint of the ocean bottom beyond the far reaches of the continental margin? Within the United States the answers to these questions interest, among others, the Department of Defense, the Interior Department, the State Department, petroleum interests, hard mineral interests, fishing interests, the scientific community, and Congress.

Interest in offshore developments is also high among the world's developing states. They have an important interest in offshore resources in terms of what they can do to the world economy, prevailing trade patterns, and the balance of payments of various states. The less developed states are, primarily, raw material exporters and the importers of capital goods and finished products. If, through undersea mineral exploration and exploitation under national control, large mineral finds are made by technologically advanced states and efficient extraction processes are developed, the dependence of these advanced states on the natural resources of the less developed states may decrease. The ability of the less developed states to earn foreign exchange could, as a direct result, decrease, and the less developed states could fall still farther behind in the development race.[7] The advanced states might well desire to favor further claims to offshore areas precisely in the hope of finding needed and vital resources and thus decreasing their dependence for strategic materials on far away and often unstable states.[8] Moreover, for some people in an advanced country such as the United States, cutting expenditures on raw material imports would be viewed as an effective means of improving a situation of an unfavorable balance of payments.

At the same time, many of the developing states are interested in establishing an area of the ocean bottom as property owned by an international authority and used for the benefit of mankind.[9] If an area is to be established under the control of some international authority, the matter of the limit of national jurisdiction will have

---

*The continental margin includes the continental shelf, continental slope, and continental rise.

to be squarely faced. Where does the area of international jurisdiction begin and where does the area of national jurisdiction end? With the present state of technology and knowledge of the oceans, it is in the areas closer to shore, the areas on the continental margin, and the areas of lesser water depths, that petroleum exploitation is most feasible and economical. Yet a developed state such as the United States may desire to establish and maintain exclusive control over the entire continental margin because of potential benefits that can accrue to it through resource exploitation.

Important issues are at stake in the debate on the continental shelf and important interests feel that there is much to gain or lose through different continental shelf policies. There is here a classic case of the sort of differences that political authorities must decide: the need to reconcile, to the extent possible, conflicting demands and to decide who gets what, what interests are served, and at whose expense.

Since the United States is a world power and leading sea power, the substantive approach of this country is of itself of great importance in the determination of the future regime of ocean space. An examination of developing American policy, however, also affords us an excellent opportunity to examine the process whereby the United States determines such policy and relevant international law positions. Politics is widely regarded as the process whereby values are distributed by public means.[10] In the domestic political sphere interest groups and individuals participate, attempting to influence favorably the outcome of political struggles. On the international level, states, international organizations, and various transnational actors participate in the political process and they too seek to influence the resolution of political differences toward their point of view. There is a tendency on the part of some to consider politics as somehow separate from the lawmaking and law administering process. A more realistic view observes that "politics and law are indissolubly intertwined."[11] In terms of the foreign policy process, it is understood that various interests attempt to influence particular decisions. Foreign policy decisions clearly emanate from the "political process" described by Roger Hilsman.[12] But what about the situation in which a state must determine not only a policy position but also a position on an international law question? Is there some difference in the way the United States makes a policy decision, that is, taking an action within the recognized options allowed by international law, and the way it makes a decision embodying either a departure from existing international law or setting a precedent in an area where there had been no existing international law? Is the difference merely one of increased participation on the part of the State Department Legal Adviser's Office in the policy-making process?

As with policy preferences, one finds that international law positions reflect a certain desired value allocation.[13] Law clearly benefits some more than others. A particular law may benefit some to a great extent, others not at all; some may even suffer detrimental effects from new law. Is it not realistic to expect that in the process whereby an international law position is determined by a government, interested parties, feeling that they stand to gain or to lose, will try to influence the nature of that legal position? One would expect that as long as international law positions affect value distribution they too would be subject to pressure group activity. It would seem that interested parties, whether governmental or nongovernmental, would not allow legal positions to be solely based upon what lawyers like to refer to as the "merits" of the case. As Carr has noted, "law is not an abstraction. It cannot be understood independently of the political foundation on which it rests and of the political interests which it serves."[14]

According to Stanley Hoffman, legal decision making is in fact a form of policy since every state wishes to turn its interests, ideas, and gains into law. International law enshrines and elevates the interests or norms which it embodies.[15] It is a fact that mere convergence of state behavior between members of the international system may exist at a given time yet there may be no rule requiring this behavior. Giving this behavior the character of an obligation of international law provides insurance against undesired changes in conduct. If, as Harold Lasswell claims, the question that concerns a political strategist is how to manage the existing system to obtain favorable outcomes and how to modify the system in order to effect future results,[16] then we are justified in considering the creation or shaping of norms of international law as useful means with which to achieve and secure policy goals.

While protection against undesired change is increased by incorporating a preference into international law, it must also be noted that freedom of future action becomes more limited. New developments and changed conditions may lead to rethinking as to the value of preferences pursued in the past and alternative preferences might take on greater importance. If, however, the first set of preferences has become embodied in international law a simple unilateral policy change does not alter the international responsibilities of a state under international law. Bringing about change in the international legal order will require multilateral action or at least international acquiescence. Typically, simple policy changes do not require international approval.

Writing on the connection between lawmaking and politics, in terms of the operation of the United States government, Professor Henkin has observed that new law is often the vector resulting from the struggle of diverse forces. In international law, too, the position of the United States will reflect the varying pressures that are brought to bear on the issue.[17] Does this, in fact, suggest the way in which an American position of an international law question is determined? If so, how do varying points of view make themselves felt? How are differences in preferences resolved?

While from the standpoint of the United States government it is the Department of State that is officially the executive department primarily responsible for the determination of American foreign policy and American positions on international law questions, actions of other executive departments can clearly have important policy and legal significance on the international level. In such a case the State Department will have to deal with the international consequences of an act or policy statement that was perhaps made because of the pressure of purely domestic political considerations and bureaucratic infighting.

In regard to American policy on the continental shelf, we will see a situation in which a department other than State, namely the Interior Department, has played a leading role in shaping the American claim to the seabed and subsoil of areas off American coasts. It was the Interior Department that provided the basic impetus that resulted in the Truman Proclamation on the Continental Shelf. From the time of the Truman Proclamation, it has been the Interior Department that has had the day to day responsibility for the administration of the American continental shelf areas. When the Geneva Convention on the Continental Shelf came into force it was left primarily to the Interior Department to apply and interpret its provisions with respect to American continental shelf areas.

The Interior Department brings to bear a special perspective peculiar to itself because of the specific interests it represents and serves. The State Department, for its part, must of necessity have a different perspective. It does not have an important fixed domestic clientele as does the Interior Department. This may account in large part, for example, for the great difficulty State usually encounters in getting from Congress what it wants in terms of foreign aid. In a certain sense the clientele of the State Department consists of foreign states and, thus, the department is to a great degree "other directed." An executive department such as Interior is, just to the contrary, "inner directed," in the sense that its programs have a very direct and immediate impact on important segments of the American public consisting of people who exert considerable influence in domestic political processes. As a result, a department such as Interior may try to satisfy its domestic political clientele and leave the foreign policy or international law implications of its policies to the State Department to handle as best it can.

To a great extent the focus of study of international law has been at the level of the international system. International law has been viewed as a result of state interaction. While this view certainly has its validity, it does not present a complete picture. It is suggested that there is also a need for the study of subsystem factors as they relate to the development of international law. Graham Allison and Morton Halperin have recently claimed that the usual approach to foreign policy visualizes international politics as involving acts of unified national governments and suggests that governmental behavior is analagous to the coordinated acts of individuals. In line with this "rational actor paradigm," analysts examine the interests and goals of a state, the alternatives available, and the costs and benefits of the different alternatives.[18] With this model of decision-making the government is conceived of as a rational unitary decision maker. Therefore, if a nation performed a particular action it must have had ends toward which the action was directed.[19] Does this model suggest the manner through which an international law position is developed? Or, on the other hand, is an international law position more the result of what Allison terms the "bureaucratic politics paradigm?"[20] In this model decisions are seen as intranational political resultants. Thus what happens is not a chosen solution to a problem but instead results from compromise, conflict, and confusion of officials with varying interests, unequal influence, and different information and perspective.

Continental shelf policy has been decided in a context of unclear international law. The problem has not been one of merely deciding between two or more policy options all fitting neatly within the existing international law framework. The need has been to take a policy position with the understanding that this position might have very significant international law ramifications. From the earliest period of consideration of the problem within the United States government, the issue has clearly been perceived to have a definite and important international law dimension.

Through an examination of the case study of American policy on the continental shelf, a number of general questions involving the relationship of lawmaking to politics can be approached. Why and how are new lines in the development of international law decided upon? How is this line of development undertaken? Who within the United States government participates in decisions of this type? What kinds of considerations are taken into account? Once a principle of law is accepted, how and by whom, within the United States government, is the new principle of law applied and interpreted? What is the effect of such application and interpretation? How are undesirable changes in the law opposed? While this study may suggest answers to such questions, additional case studies of this type will be needed in developing generalizations.

NOTES

1. On the general concept of the continental shelf, see M. W. Mouton, The Continental Shelf (The Hague: Martinus Nijhoff, 1952), ch. 1.

2. C. H. M. Waldock, "The Legal Basis of Claims to the Continental Shelf," Transactions of the Grotius Society 36 (1950): 115.

3. General Assembly Resolution 2750 C (XXV), December 17. 1970.

4. National Petroleum Council, Petroleum Resources Under the Ocean Floor (Washington: NPC, 1969) p. 7.

5. Press Release, United States Information Service, Geneva, Switzerland, statement of Dr. Vincent McKelvey, United States Geological Survey, before subcommittee I of the Enlarged United Nations Committee on the Peaceful Uses of the Seabed and the Ocean Floor Beyond the Limits of National Jurisdiction, dated August 4, 1971.

6. See, for example, the testimony of Professors Louis Henkin and Carl A. Auerbach before the Special Subcommittee on the Outer Continental Shelf of the Senate Interior Committee, "Outer Continental Shelf," 91st Congress, 1st and 2d sessions, pp. 183-184 and 196 respectively.

7. African states, for example, produce about 81 percent of the world's cobalt supply and about 50 percent of the world's manganese. Countries like Zambia, Zaire, Gabon, and Ghana need the earnings derived from the export of these resources. Conceivably such states may be reluctant to see the jurisdictional questions concerning offshore areas resolved. On manganese nodules see David B. Brooke, "Deep Sea Manganese Nodules: From Scientific Phenomenon to World Resources," Proceedings of the Second Annual Conference of the Law of the Sea Institute (Kingston: University of Rhode Island, 1968) pp. 32-41.

8. This consideration remains valid, even though from a purely economic point of view the undersea source would be more expensive to exploit. Walter Levy, "Oil Power," Foreign Affairs 49 (July 1971): 652-668.

9. This is evidenced by the support of the less developed states for General Assembly Resolution 2749 (XXV), the Declaration of Principles Governing the Seabed and the Ocean Floor, and the Subsoil Thereof, Beyond the Limits of National Jurisdiction. This resolution was approved on December 17, 1970.

10. Harold Lasswell, Politics: Who Gets What, When, How (New York: McGraw-Hill Book Company, 1936).

11. E. H. Carr, The Twenty Years Crisis, 1919-1939 (New York: Harper and Row, 1964), p. 177.

12. According to Hilsman the political process involves reconciliation of diversity of goals and policies, competing groups identified with varying alternatives, and consideration of the relevant power of these groups. Power considerations in this process are as important as the wisdom of the arguments used in support of different policies. Hilsman, "The Foreign-Policy Consensus: An Interim Research Report," Journal of Conflict Resolution 3 (1959): 365. The debate on continental shelf policy, both domestically and internationally, has these characteristics.

13. This explains why, for example, the less developed states want a greater role in the creation and development of international law. On this point see Jorge Castaneda, "Underdeveloped Nations and the Development of International Law," International Organization 15 (1961): 38-48.

14. Carr, op. cit., p. 179. See also Kaplan and Katzenbach, The Political Foundations of International Law (New York: John Wiley, 1961).

15. Hoffman, "The Study of International Law and the Theory of International Relations," in his The State of War (New York: Praeger Publishers, 1965), pp. 123-133.

16. Lasswell, op. cit., (Meridian edition, 1958), pp. 205-206.

17. Henkin, "International Law and 'the Interests': The Law of the Seabed," American Journal of International Law 63 (1969): 505.

18. Allison and Halperin, "Bureaucratic Politics: A Paradigm and Some Policy Implications," World Politics 24 (1972): 40-79 at p. 41.

19. Allison, Essence of Decision: Explaining the Cuban Missile Crisis (Boston: Little, Brown and Company, 1971) pp. 32-35.

20. Ibid., pp. 162-184.

CHAPTER

2

EVENTS LEADING TO THE 1945 TRUMAN PROCLAMATIONS

Political concern with the continental shelf began this century in connection with the matter of fisheries. The fishing industry was changing, newer equipment was being used, and the size of the catches was increasing. Change, however, did not benefit the fishermen of all states equally. At the beginning of the twentieth century some fishing states were becoming disturbed by the pressure of foreign long-distance fishing vessels operating off their coasts. There was some dissatisfaction with the traditional three-mile limit of exclusive coastal fishing and some fishermen suggested that national jurisdiction be extended over all the waters of the continental shelf.[1]

OFFSHORE FISHING

In the middle to late 1930s, the continental shelf finally came to the attention of American policy makers as a result of concern with Japanese fishing activities off the coast of the United States. The Alaskan salmon industry feared depletion of the salmon stock because the Japanese took the salmon from the shallow waters over the continental shelf as they came from the ocean to inland waters to spawn. This Japanese "invasion" of salmon fisheries in the Bristol Bay area during the 1936-1938 period created tremendous resentment in the United States against the Japanese.[2] The persistence of these fishermen led to strong political pressure on the United States government to take action to limit the right of foreign nationals to fish in the waters above the continental shelf of Alaska.

In 1937 Senator Bone introduced a bill which would have extended the jurisdiction of the United States offshore to a depth of 100 fathoms "for the necessary protection and preservation of the salmon fishery."[3] The State Department, while claiming sympathy for the purposes of the bill, pointed out that this bill, if made into law, "would be without legal precedent." It noted that "the dangers of international controversies and dissensions inherent in such a proposal are at once apparent." The additional observation was made that if other states made similar claims for their coastal areas, the United States government would find it very difficult to object to the application of the principle against our nationals and vessels.[4] According to the Department of State, a solution should be sought through negotiations with Japan rather than through any unilateral action. In deference to this view the Senate Commerce Committee, chaired by Senator Copeland, held no hearings on the bill and nothing ever came of it.

While the State Department negotiated with Japan, Delegate Dimond of Alaska introduced a bill in the House of Representatives in February of 1938 which called upon foreign nationals to refrain "from engaging in salmon fishing in the waters adjacent to the coast of Alaska as far as the outer limit of the continental shelf." The bill would empower the President to establish salmon fishery law enforcement areas; enforcement powers were also to be granted to the federal government giving it the right to confiscate illegal catches.[5]

To counter the public pressure for action, such as that embodied in the Dimond bill, which might have dangerous consequences for Japanese-American relations, and to show that it was doing something to defend American interests, the State Department made public a communication it had sent to the Japanese government, which noted the great anxiety within the United States over the salmon fishery problem. The statement's import was that some agreement had to be reached or there would be no alternative but to take unilateral action.[6]

A person involved in the events of this period has claimed that Secretary of State Cordell Hull gave personal assurances to him and to Senator Schwellenbach that measures would be undertaken to stop the Japanese intrusion. To secure immediate relief, negotiations were opened with the Japanese; the desire to prevent future repetitions developed into the Truman Proclamation on Fisheries.[7] In 1938 the United States, in fact, did secure a temporary ban from the Japanese government against continued use by its nationals of the Alaska salmon fisheries. With the achievement of a temporary settlement with Japan relative to the west coast fisheries, the attention of the State Department shifted away from fishery matters to more urgent and pressing problems.

PETROLEUM EXPLOITATION

As early as 1894 oil had been extracted from shallow submerged land off the coast of California. The existing technology was simple and for some time there was no question at all of operating beyond the three-mile limit of territorial waters. In 1918, however, a private citizen wrote a letter to the Department of State inquiring how he might acquire property rights to petroleum deposits that he believed to exist in land beneath the Gulf of Mexico about 40 miles from the closest land but in water of less than 100 feet. The State Department response noted that "the United States has no jurisdiction over the ocean bottom of the Gulf of Mexico beyond the territorial waters adjacent to the coast."[8]

In 1938 the Independent Exploration Company sought to be employed by the Department of the Interior to perform geophysical exploration work beyond the three-mile limit in the Gulf of Mexico to ascertain the presence of oil. This request caused considerable concern within the Interior Department. In considering a reply to this inquiry, the Director of the Geological Survey ventured to give his own opinion that "Under international law, the oceans beyond the three-mile limit are the common property of all nations, and no nation has exclusive jurisdiction over them."[9] When consulted, the Solicitor of the Interior Department suggested caution since he knew of no legal basis upon which to grant exploration rights beyond the three-mile limit.[10]

The problem of rights to the submerged land beyond the three-mile limit came to the attention of President Roosevelt soon after the request of the Independent Exploration Company. In the latter half of 1938, Roosevelt sent a proposal to the Department of the Interior in which he suggested that he "issue an Executive order setting up naval oil reserves on the coast beginning with the shore line and extending halfway across the ocean. . . ." Interior Solicitor Margold indicated, however, that in accordance with international law the seabed outside of territorial waters was owned by no one and free to all. He claimed that prior to occupation there was no basis under which such land could be considered public land of the United States.[11] President Roosevelt, dismayed at the timid response with which his bold proposal was greeted, maintained that American jurisdiction could be exercised far out into the ocean. Responding to the doubts about the international legality of such a claim, he wryly noted:

> I recognize that new principles of international law might have to be asserted but such principles would not in effect be wholly new, because they would be based on the consideration that inventive genius has moved jurisdiction out to sea to the limit of inventive genius.[12]

As the Interior Department began to give its attention to the problem of the submerged lands beyond the three-mile limit, the

United States Navy was expressing its fears that the supply of oil available to the United States would be inadequate unless some substantial new finds were made. Captain Stuart, the director of the Naval Petroleum Service, claimed that the submerged land under American territorial waters, that is within the three-mile limit, held substantial promise as a new source of petroleum.[13] While Captain Stuart was concerned only with the land beneath American territorial waters, he helped to focus American attention on the possibility of seeking more oil from offshore areas generally. When asked by a congressional committee about oil operations beyond the three-mile limit, then Assistant Solicitor of the Interior Department Harry Edelstein replied that he did not know about the legal situation beyond the limit of territorial waters.[14] This was a subject, however, to which the Interior Department was to give increasing thought.

By 1939 several congressional committees were investigating the whole matter of offshore oil exploitation. At more than one of these hearings attention turned to the question of ownership or jurisdiction over submerged lands beyond the three-mile limit. It was clear that ownership of the land within the three-mile limit was American (whether it belonged to the states or to the federal government was quite another matter). No claim, however, was made relative to the submerged land beyond territorial waters. A senior attorney of the office of the Judge Advocate General of the United States Navy commented at the time that an American claim "would not be a legal exercise of rights so far as other nations are concerned."[15] In spite of such testimony, Senator Holman of Oregon strongly urged that the United States undertake control of offshore lands to the edge of the continental shelf.[16]

THE SHAPING OF THE TRUMAN PROCLAMATIONS

With the outbreak of World War II, American concern with the problem of coastal fisheries and petroleum rights beyond the three-mile limit decreased. While it was true that the war effort required tremendous quantities of petroleum and petroleum products, little energy was expended in the direction of exploiting possible oil supplies in the continental shelf areas off American coasts. In the early 1940s the technology of oil extraction from submerged land was in the process of development and was expensive relative to the cost of onshore operations. It was not until the period following World War II that offshore petroleum exploitation was significantly developed.

Though general interest did lessen during the war years, positions were being formulated within the United States government for the postwar period. For some time Secretary of the Interior Ickes had been aware of the probability of oil deposits in areas under the open seas off California and the Gulf coast because of studies conducted by the Interior Department's Geological Survey.[17] Apparently, though, it was a memorandum in May of 1943 from the Interior Department's General Land Office that caused Secretary Ickes to take a determined lead on the matter of the continental shelf. The memorandum noted the need for natural resources and the belief of scientists that such resources were to be found on the continental shelf. It indicated not only that there were important petroleum resources to be found there but also that the continental shelf was important as a breeding ground for fish. In addition, it alluded to security implications by pointing out that the shelf could also serve as a hibernating place for submarines. In the light of these considerations, it was suggested that the United States substitute a line of 100 or 150 miles from its shore for the three-mile limit of territorial waters.[18]

A short time after Secretary Ickes had received this memorandum, he brought the matter to the attention of President Roosevelt, urging him to take steps so as to avail the United States of the riches in this submerged land and in the waters above it. Ickes noted that to do this, new concepts of territorial limits beyond three miles would have to be formulated.[19]

The impact of this letter seems to have been considerable. In July of 1943, Assistant Secretary of State Long informed Leon Sturgeon, who had been working on fisheries policy within the State Department for some time, that the President wanted consideration given to mineral resources as well as to fisheries in any extension of American jurisdiction offshore.[20] From this time on the State Department worked together with the Interior Department. The main concern of the State Department, however, remained the matter of fisheries. The de facto division of labor became clearer when, in July of 1944, it was agreed to separate the statement on fisheries from that on the continental shelf. The statement on the continental shelf was, in fact, prepared by the Department of the Interior.[21]

Once Secretary Ickes became convinced of the potential value of the continental shelf to the United States he emerged as the strong champion of extended American rights to offshore areas, fighting for absolute sovereignty over the shelf and the waters above it. Writing soon after the conclusion of the war and the issuance of the Truman Proclamations on the continental shelf and on fisheries, Ickes indicated his disappointment that his position was not adopted. According to Ickes his plan was whittled down by the lawyers.[22] The lawyers to whom he refers were in the State Department Legal Adviser's Office and they were the individuals who were chiefly responsible for the modifications forced upon the Interior Department. According to one of these attorneys, the objections to a United States claim of sovereignty over the entire continental shelf and the waters above it were varied. First, it would be difficult to obtain acceptance from other states of a claim to territorial waters far exceeding the generally observed three-mile limit. Second, such a claim by the United States would encourage other states to make similar claims of their own, thus leading to interference with normal American fishing operations in places such as Mexico and Cuba (shrimp) and the west coast Latin American states (tuna). Third, wider territorial seas would make it more difficult for the United States Navy since unfriendly ships could take refuge in neutral territorial waters of correspondingly wide extent. Fourth, there was really no need to claim sovereignty when all that was called for was control over the resources of the seabed of the continental shelf.[23]

In suggesting that the United States claim sovereignty over the continental shelf and the epicontinental waters, Ickes, as Secretary of the Interior, was thinking primarily in terms of United States' claims to the submerged land, waters, and air space off American coasts; the Department of State, understandably, was equally concerned with the right of access for American vessels, fishermen, and aircraft in the vicinity of the coasts of other countries. In light of the position and responsibilities of the State Department it is not at all surprising that State should stress these considerations and raise the question clearly implied by point four above; namely, why claim more than the United States absolutely needed? Though modified by attorneys of the Legal Adviser's Office, there remained strong opposition within the State Department to the claim, due to its unilateral character.[24]

By December of 1944, however, agreement was reached between the Departments of State and Interior on the continental shelf and fisheries proclamations.[25] Though he was not totally pleased, Secretary Ickes fought hard for the immediate announcement of these proclamations and to this end attempted to bypass the regular channels. As delays occurred time and time again, Ickes became increasingly impatient. In March of 1945 he spoke with Under Secretary of State Grew wanting to know what had happened to the memorandum agreed upon by State and Interior, which was supposed to be sent to the President for approval. In no uncertain terms Ickes stated that he wanted the matter "pushed forward."[26] Throughout this period there seemed to be the fear on the part of Interior that unless the new continental shelf proclamation were quickly announced a policy reversal would occur. As previously noted there was, in fact, strong opposition within the State Department to what Ickes believed to be a rather modest claim. In any case the matter was, indeed, brought to the attention of President Roosevelt and was approved by him on March 31, 1945. It was agreed, however, not to publicize these proclamations until informal consultations had been held with several states.[27]

On April 12, 1945 President Roosevelt died and this development contributed to further delay in the issuance of the presidential proclamations. President Truman needed time to become accustomed to his new responsibilities and had to be informed of many issues. Presidential approval was needed anew. In a detailed memorandum the State Department informed the new President as to the necessity for approving the two pending proclamations.[28]

President Truman gave his verbal assent in early May. On May 9 and 12, British and Soviet officials were notified of the proposed proclamations and it was suggested to them that their governments take similar action concurrently with the United States.[29] Other states which were told of the proclamations were Canada, Mexico, the Netherlands, Norway, France, Iceland, Denmark, Cuba, and Portugal.[30] The United States was thus sampling possible reactions to the American initiatives prior to making them public.

While the State Department was waiting to see if any diplomatic protest would result, Interior Secretary Ickes was fuming at the continued delay. In a bitter letter to Secretary of State Stettinius, Ickes commented:

> I should have supposed that the time for obstruction was over when two Presidents, to say nothing of the Secretary of State, have made a decision. I am sure that you will not think it unreasonable of me to ask the proposed proclamations and Executive orders be submitted to the President for signature as soon as possible.[31]

Unfortunately for the impatient Ickes, the situation was further complicated by the fact that on July 2, 1945 James Byrnes replaced Stettinius as Secretary of State. In reply to the above letter of Ickes, Byrnes noted that he had just taken office and "as I am sure you will understand, this matter is new to me."[32] Thus the delay could be expected to continue. In his response to Ickes, Byrnes had suggested that before taking final action a meeting should be arranged with the members of the Senate Foreign Relations and Public Lands Committees. Ickes agreed to this and during the summer these senators were briefed. Some senators thought that the continental shelf proclamation would prejudice the claims of the states to submerged lands vis-à-vis the federal government. To allay these fears a sentence was added to Executive Order 9633 indicating that neither the proclamation nor this executive order would affect the question of ownership of the tidelands.[33] On September 28, 1945 the proclamations were at last released.

Thus we find that concern with the continental shelf, first seen in relation to fisheries, broadened so as to include nonliving resources.

Within the United States government, the Interior Department, under the leadership of Harold Ickes, became the dominant force behind the claim to the continental shelf. The claim of the United States embodied in the Truman Proclamation was obviously not sufficiently all-encompassing from the point of view of the Interior Department, but it was believed to be an initial step in the right direction. Opposition from the State Department had forced Interior to settle for less than it had originally desired.

NOTES

1. L. Leonard, International Regulation of Fisheries (Concord: Rumford Press, 1944) p. 35 and Edwin J. Cosford, "The Continental Shelf 1910-1945," McGill Law Journal 4 (1958): 246.

2. Joseph W. Bingham, "The Continental Shelf and the Marginal Belt," American Journal of International Law 40 (1946): 175.

3. Congressional Record, June 18, 1937, p. 5953.

4. Letter of Secretary of State Hull to Senator Copeland dated August 4, 1937, reproduced in Foreign Relations of the United States, 1937 (Washington: Government Printing Office, 1954) 4: 755-757.

5. H. R. 8344, 75th Congress, 3rd session.

6. State Department Press Release of March 25, 1938 in United States Department of State Press Releases 18: 414.

7. Bingham, op. cit., pp. 175-176 and Wilbert Chapman, "The United States Fish Industry and the 1958 and 1960 United Nations Conference on the Law of the Sea," in Proceedings of the Third Annual Conference of the Law of the Sea Institute (Kingston: University of Rhode Island, 1969), p. 36.

8. Green Hackworth, Digest of International Law (Washington: Government Printing Office, 1941) 2: 680.

9. Memorandum from the Director of the Geological Survey to the First Assistant Secretary of the Interior, dated May 10, 1938. In the files of the Interior Department, file no. 2 208, pt. 2.

10. Memorandum from Solicitor Margold to the First Assistant Secretary of the Interior, dated May 19, 1938. Ibid.

11. Memorandum from Solicitor Margold to the First Assistant Secretary of the Interior, dated September 16, 1938. Ibid., pt. 3.

12. Memorandum from President Roosevelt to the Secretary of the Interior, dated July 1, 1939. Ibid.

13. Hearings on S. J. Resolution 208, United States House of Representatives, Committee on the Judiciary, 75th Congress, 3rd session, "Title to Submerged Oil Lands," p. 36.

14. Ibid., p. 66.

15. House of Representatives, Committee on the Judiciary, 76th Congress, 1st session, "Submerged Oil Lands," p. 44.

16. United States Senate, Committee on Public Lands and Surveys, "Title to Submerged Lands," 76th Congress, 1st session, p. 16. See also the views expressed in American Bar Association, Section of International and Comparative Law, "Title to Petroleum Beneath the Bed of the Sea," in Selected Papers and Reports, 1940, pp. 44-46.

17. See, for example, the Memorandum for the Secretary of the Interior from the Geological Survey, dated May 23, 1940, in the files of the Interior Department.

18. Memorandum from the General Land Office, Department of the Interior, to Interior Secretary Ickes, dated May 28, 1943. In the files of the Interior Department, file no. 2 208, pt. 4.

19. Letter from Secretary Ickes to President Roosevelt, dated June 5, 1943. In the files of the Department of State, 811.0145/11-2844.

20. Letter of Assistant Secretary of State Long to Leo Sturgeon, dated July 6, 1943. In Breckenridge Long papers, Library of Congress.

21. Memorandum on the meeting in Assistant Secretary of State Long's Office, July 25, 1944. State Department files 811.0145/7-2644.

22. This information is contained in the rough draft of an article on the continental shelf written by Secretary Ickes and entitled "Our Submerged 20 Per Cent." It was dated October 25, 1945 and was found in the files of the office of the Interior Solicitor.

23. This information was contained in a letter to the author.

24. Memorandum for the Secretary of the Interior from Michael W. Straus, Assistant Secretary, dated August 12, 1944 and found in the files of the Interior Department.

25. Letter of Secretary Ickes to the Secretary of State, dated December 19, 1944, in the files of the State Department, 811.0145/11-2844.

26. Memorandum of Undersecretary of State Grew, dated March 23, 1945, in the files of the State Department, 811.0145/3-2345.

27. Memorandum from the President, in the files of the State Department, 811.0145/3-3145.

28. Memorandum of the Department of State, "Explanatory Statement on the Proper Utilization and Development of Natural Resources of the Subsoil and Sea-bed of the Continental Shelf," dated April 26, 1945, in the files of the State Department, 811.0145/4-2645.

29. See the memorandums of E. H. Dooman, Special Assistant to Assistant Secretary of State Dunn on his conversations with British and Soviet officials on May 9 and May 12, 1945. Foreign Relations of the United States, 1945 2: 1504-1509.

30. Memorandum of Dooman, dated June 15, 1945, ibid., pp. 1510-1511.

31. Letter of June 28, 1945, ibid., p. 1516.

32. Letter of July 5, 1945, ibid., pp. 1519-1520.

33. See the memorandum of Acting Interior Secretary Fortas to the Acting Secretary of State, dated September 17, 1945 and the letter from the Acting Secretary of State to Senator Connally, dated September 18, 1945. In the files of the State Department, 811.0145/ 9-1745 and /9-1845.

CHAPTER

3

THE TRUMAN PROCLAMATION TO THE GENEVA CONFERENCE ON THE LAW OF THE SEA

Important international law considerations were pondered within the United States government in the process of establishing an American claim to rights in offshore areas beyond the limits of territorial waters. It was recognized that there was little precedent for the American action; this point was underscored by the fact that the continental shelf proclamation made no reference to any legal rationale for the claim it contained. Instead justification within the proclamation was based on the belief that American action was desirable, necessary, reasonable, just, and important to American self-defense. While technological development had made it physically possible to exploit petroleum deposits beyond the limits of territorial waters, international law had not demonstrated corresponding growth. What was needed was the "inventive genius" that President Roosevelt had suggested.

The collective inventive genius of the United States government, as seen in the resulting Truman Proclamation on the Continental Shelf, advanced a claim that was substantively tailor-made to fit American needs. The desired mineral resources of the continental shelf were to appertain exclusively to the United States and to be subject to its laws. Thus American industry was to be provided the legal stability needed to encourage the exploitation of such minerals. At the same time, the waters above the continental shelf were to keep their legal character as high seas thereby allowing unrestricted navigation and overflight. Assuming that other states made similar claims on their own behalf, there would be no detrimental effect upon the uses of the seas so important to the United States Navy. Moreover, the rights of American fishermen who operated off the coasts, but beyond territorial waters of other states, would not be violated.

When a new claim is made in international law, however, there is no assurance that it will be accepted or that it will be accepted in the form originally advocated. It has thus far been shown how an idea developed within the United States government and how that conception manifested itself as an international law claim. Now we focus upon the different forums in which this claim was considered and the attempts to shape and control its growth and development so as to keep it compatible with perceived American interests. In this regard we will examine the claims of foreign states, American domestic developments, the actions of the inter-American system, and the impact of the International Law Commission of the United Nations.

AN ANALYSIS OF THE TRUMAN PROCLAMATION

With the issuance of the Truman Proclamation on the Continental Shelf on September 28, 1945, the United States had made a claim vis-à-vis the international community. Specifically, the United states government made it known that it "regards the natural resources of the subsoil and sea bed of the continental shelf beneath the high seas but contiguous to the coasts of the United States as appertaining to the United States, subject to its jurisdiction and control." The claim embodied in this proclamation involved the assertion of a new principle of international law. This being so, it was necessary for the policy makers to formulate a position that would both secure American objectives and minimize the possibility of objections by foreign states.

While the United States wanted exclusive jurisdiction over the petroleum deposits in the continental shelf, it sought to maintain certian existing freedoms in the waters beyond the traditional three-mile limit. The Truman Proclamation specifically stated that the "character as high seas of the waters above the continental shelf and the right to free and unimpeded navigation are in no way thus affected."

The United States claimed only that the natural resources of the subsoil and seabed of the continental shelf were subject to its "jurisdiction and control." Jurisdiction and control was not claimed over the continental shelf generally in the Truman Proclamation. An individual who participated in the policy making of this period suggests that jurisdiction and control over the natural resources of the continental shelf, rather than over the subsoil and seabed, was asserted in order to prevent any inference that control over the resources of the seabed carried with it control over the superjacent waters, free-swimming fish therein, or the air space above.[1]

For the same reason "sovereignty" was not claimed over the resources of the seabed. As seen earlier, Interior Secretary Ickes had sought a United States declaration of sovereignty over a wide offshore area. The term "sovereignty," however, was rejected by officials of the State Department Legal Adviser's Office because it was felt that a claim of sovereignty was unnecessary for the limited purpose of petroleum exploitation. It was believed that such a claim would lead to undesirable consequences. Jack Tate, who served in the office of the Legal Adviser, has pointed out that a claim of sovereignty might well have been interpreted by other states as affecting the character of the waters and of the air space above the seabed despite disclaimers to the contrary.[2] A more limited claim would still provide sufficient basis for the orderly development of offshore natural resources.[3]

As a means of further emphasizing the distinction that American policy makers wanted to draw between the seabed and the waters above, a separate proclamation was issued on the subject of fishery policy. This second proclamation, issued concurrently with the continental shelf proclamation, would empower the United States to establish fishery conservation zones in areas of the high seas contiguous to its coasts. In such areas, fished in only by American nationals, the United States would by itself establish regulations. Where foreign fishermen also operated, such conservation zones would be established under international agreement with foreign states. In this proclamation the United States recognized the right of other states to take similar action off their shores if recognition were given to existing fishing interests of American nationals. The fisheries proclamation was a conservation claim. As Wilbert Chapman, Special Assistant to the Undersecretary of State, noted, the proclamation did not extend American sovereignty beyond the three-mile limit or attempt to exclude fishermen of any nationality from any fishery.[4] This proclamation stressed, as had the continental shelf declaration, that the character as high seas of these areas and the right of free navigation were in no way affected. Despite the intentions of the United States government, however, a number of Latin American states understood or interpreted these proclamations as making a claim to sovereignty over offshore areas and began asserting wide ranging claims of their own.

Unnecessary confusion seems to have been generated by the fact of the simultaneous issuance of the two proclamations. The question naturally arises: how are the two proclamations related to one another? With the benefit of hindsight, Jack Tate has observed that the proclamations should have been issued separately; according to him they related to two entirely different matters and were based upon two different theories.[5]

While the vertical dimension of the claim made by the United States was seemingly apparent from the text of the continental shelf proclamation itself, the horizontal extent was not so clear. No elaboration of the breadth of the continental shelf area appears in the proclamation. A White House press release of the same date as the proclamation did, however, indicate that the proclamation applied to the submerged land out to a water depth of 100 fathoms or 600 feet.[6] The horizontal extent of the claim made by American policy makers was not perceived to be open ended.

The leadership of the United States Interior Department on the continental shelf issue resulted in bureaucratic rewards for that department. President Truman, in an executive order issued jointly with the two presidential proclamations, mandated that the natural resources of the shelf be placed under the jurisdiction and control of the Secretary of the Interior.[7] Despite the demonstrated concern of the State Department with regard to the wording that was employed in the Truman Proclamation, Secretary Ickes was to write, after his department had been given administrative control over the continental shelf areas, that the "two presidential proclamations assert our sovereignty over the mineral resources of this ground and our jurisdiction over the fishery resources of the high seas contiguous to our lands."[8] In an article written in 1946, Ickes, still Interior Secretary, wrote that by the Truman Proclamations, we, the United States "expanded, in effect, our frontiers."[9] It is understandable that statements of this nature by high-ranking American officials could well cause confusion as to what, exactly, the United States was claiming.

In analyzing the import of the Truman Proclamations we must consider not only the substantive claims made but also the manner in which the claims were made. Despite the fact that several states had been informed in advance, the significant point is that the proclamations were unilateral in nature. This was later to be regretted. Former State Department Legal Advisor John Stevenson has observed that the continental shelf proclamation has been repeatedly used as a justification by other states for their own unilateral claims to the waters above the shelf as well as to the seabed itself. Defense Department spokesmen have made the same point.[10]

FOREIGN CLAIMS TO OFFSHORE AREAS

Shortly after the issuance of the Truman Proclamations, other states began staking out unilateral claims of their own to offshore areas. In a declaration by the President of Mexico on October 29, 1945, Mexico claimed possession of the entire continental shelf adjacent to its coasts and all of its natural resources. Additionally, Mexico was to undertake the supervision, utilization, and control of fishery zones that were necessary for conservation.[11]

Insofar as Mexico was making a claim to the mineral resources of its continental shelf, the United States was not perturbed. This was so in light of the fact that the Mexican declaration indicated that the character as high seas of the waters above the shelf and the right to free navigation was to be respected, and that the continental shelf was delimited by the 200 meter isobath. Thus, the limit of the shelf area claimed by Mexico was very similar to the 100 fathom limit suggested by United States authorities.[12]

American concern focused upon the possible claim to fishery resources beyond the limits of Mexican territorial waters. The Mexican declaration pointed to the foreign exploitation of fish in the waters above the continental shelf, indicating that it was indispensible for Mexico to extend its control and supervision to fishery breeding grounds irrespective of their distance from the coast.[13] In an attempt both to encourage the Mexican government to clarify its intentions and to make the position of the United States government clear, the United States specifically noted that the Truman Proclamation on Fisheries respected the fishing rights of other states.[14] In this manner Mexico was being informed that it could not unilaterally bar American fishermen from operating over the Mexican continental shelf beyond territorial waters where they had fished for some time.

The Mexican declaration with respect to fisheries, thus, was interpreted by the United States as a conservation claim rather than as a claim to exclusive fishing rights. Anticipating the adoption of measures to make the declaration effective, the United States noted that such rules should be founded upon "scientific findings showing the need for such measures from the standpoint of conservation" and developed through "cooperative arrangements with the United States in so far as the rights or interests of the United States or its nationals may be affected."[15] Mexican actions should be, to the extent that they affect American rights, neither arbitrary nor unilateral. American recognition of Mexican fishing zones was to be contingent upon these considerations.

Whereas the continental shelf declaration of Mexico could be interpreted as being consistent with the United States position as expressed in the Truman Proclamations, the stance taken by the government of Argentina was clearly out of line with American policy. In a presidential decree of October 11, 1946, Argentina claimed the entire continental shelf and epicontinental sea off its coast.[16] Either misunderstanding or misusing the precedents set by the governments of the United States and Mexico, Argentina stated that these two states had "issued declarations asserting the sovereignty of each of the two countries over the respective peripheral epicontinental seas and continental shelves."[17] Consequently, the Argentine government claimed sovereign power not only over the continental shelf but also over the waters above the shelf. While free navigation was explicitly said to be unaffected in the area of the epicontinental seas, the claim of sovereign power suggested that a change could occur in subsequent declarations. Further, the absence of any reference to fishing rights indicated that fish in the epicontinental waters could be regarded as exclusively an Argentine resource. The claim of Argentina had surpassed that of the United States.

Accordingly, the United States formally protested the Argentine decree. To have made no protest would have been viewed as acquiescence in and acceptance of the Argentine claim. Argentina was informed that the principles underlying its declaration differed in large measure from those of the United States proclamations and that they appeared to vary with generally accepted principles of international law. The United States government, therefore, put Argentina on notice that it reserved its rights and interests in the face of the Argentine decree and of any measures designed to give it effect.[18]

Soon after the claim of Argentina there followed the claims of Chile,[19] Peru,[20] and Costa Rica.[21] All three of these states made references in their proclamations to the preceding claims of the United States, Mexico, and Argentina. Chile, Peru, and Costa Rica declared their national sovereignty over the continental shelf to whatever depth of water the shelf might extend. Additionally, all these states made claims to the epicontinental seas and declared fishery zones to a distance of 200 miles from their coasts.

In the same terms used to protest the claim of Argentina, the claims of Peru and Chile were also protested by the United States.[22] For some unknown reason the United States apparently did not protest the decree of Costa Rica.[23]

In September of 1950, El Salvador adopted a new constitution, which stated in article seven that its territory included the adjacent seas to a distance of 200 miles and the corresponding air space and continental shelf.[24] In a note to the government of El Salvador, the United States pointed out that sovereignty over coastal waters was limited to the narrow belt of territorial seas recognized under international law. It was observed that if the provisions of article seven were executed, the exclusive jurisdiction and control of El Salvador would extend into wide areas of the high seas. Consequently, the United States would not consider its nationals, ships, or aircraft as being subject to the stipulations of that article or to any measures designed to give it effect.[25]

In 1952 Chile, Peru, and Ecuador jointly proclaimed that they possessed sovereignty and jurisdiction over the adjacent seas to an extent not less than 200 nautical miles from their coasts.[26] Two years later these states further agreed to consult with one another to better defend the principle of national sovereignty over this extensive maritime zone.[27] Nevertheless, whether stated unilaterally or in unison, the United States remained vehemently opposed to claims of sovereignty beyond the three-mile limit.[28]

DEVELOPMENTS IN AMERICAN DOMESTIC POLITICS

The Submerged Lands Act

While other nation-states were declaring claims of their own to continental shelf areas, the United States was seized with the important and controversial political problem of the "tidelands." The use of the term "tidelands" is misleading; what actually was at stake was the ownership of coastal submerged lands adjacent to the states within the territorial waters of the United States. The question that arose in the United States, with its federal system of government, was: who owns this land? Uncertainty as to the answer to this question discouraged commercial ventures on the continental shelf. For a long time it seemed to have been the opinion of even federal officials that this offshore land was owned by the states.[29] Pressure to assert federal jurisdiction over this area initially came from Interior Secretary Ickes.[30] In the case of the <u>United States v. California</u>, the issue of federal versus state rights in the offshore lands was dealt with, and the Supreme Court ruled in favor of the federal government.[31]

Very soon after this decision, the federal government received applications for offshore leases in accordance with the Mineral Leasing Act of February 25, 1920. In August of 1947, the Solicitor of the Interior Department and the United States Attorney General held, however, that this act did not apply to submerged continental shelf lands.[32] Thus, applications for offshore leases were denied by the Interior Department. The situation at this point was that only the federal government could lease offshore lands but the federal government had no existing statutory authority to do so. According to former Interior Secretary Chapman, President Truman recognized that a program for the development of these lands was needed and he attempted, without success, to get Congress to enact legislation to this end.[33]

Though the Supreme Court had ruled in favor of the federal government in the California case, the issue of the ownership of the tidelands was far from resolved. Great political pressure was brought to bear in Congress to recognize the authority of the states to control and benefit from the natural resources of the tidelands. This pressure eventually led to the passage of the Submerged Lands Act of 1953, by which the federal government surrendered all its rights in the offshore lands within certain geographical limits and reconfirmed its rights to the submerged lands beyond these stipulated limits.[34]

The debate on the tidelands raised several important points. First, the fact was reiterated that petroleum was a commodity of very substantial significance to the security and well-being of the United States and that existing domestic deposits of petroleum were considered too limited. Defense Secretary Charles Wilson informed Congress that in the past ten years the United States had changed from a net exporter of petroleum to a net importer and emphasized the need for development of oil reserves wherever they might be located.[35] High government officials as well as executives of the oil industry were anxious to have the continental shelf areas of the United States developed.

Second, meaningful questions were considered relating to the extent of the submerged areas that were to be given (or returned) to the states. Insofar as the states were to receive the submerged land within the three-mile limit, no international complications were foreseen. A problem was envisioned by some, though, in connection with the claims of Florida and Texas to a line of three marine leagues off their coasts as their outer boundary. The bill, which became the Submerged Lands Act, stated that the term "boundaries" as used therein was to mean the boundaries "as they existed at the time such State became a member of the Union . . ." If recognized by the federal government, the territorial waters of these two states would extend beyond those claimed by the United States. The validity of these claims, however, had not been uncritically accepted by the federal government. In a letter to Senator Connally of Texas, James Webb, the Undersecretary of State, replied to certain questions posed by the senator pertaining to the right of Mexican fishermen to operate beyond the three-mile limit but within the three-league boundary claimed by Texas. In part he observed that the United States claims territorial waters of three miles and, consequently, does not recognize the Texas claim as binding upon Mexico or Mexican nationals.[36]

The possible recognition by Congress of the three-league claims of Florida and Texas was a source of concern to American fishery interests, who emphasized the danger to their industry that would result from such an action. This recognition, it was feared, would be used as a precedent by other nation-states on which to build even more extensive claims to what had been traditionally regarded as high seas.[37]

The State Department shared these fears concerning the claims of Florida and Texas. According to this department, recognition by Congress of the claims would be a modification of the concept of the three-mile territorial sea and would be in conflict with international law. Abandonment of the three-mile limit by the United States would make it impossible to protest claims of foreign states to greater breadths of territorial waters, thus further limiting areas in which American fishermen could operate.[38]

It is interesting to note that Interior Secretary Douglas McKay was not apparently bothered by this problem. In testimony before the Senate Interior Committee he expressed his belief, and presumably that of his department, that the national interest would best be served by "restoring to the various states [of the United States] the coastal offshore lands to the limits of the line marked by the historical boundaries of each of the respective States."[39] It was made clear that by the term "historic boundaries" he meant the three-mile limit or three-league limit, whichever it was.

Despite the misgivings expressed by the State Department and representatives of the American fishing industry, the Submerged Lands Act, as finally approved by Congress and signed into law by President Eisenhower, defined "boundaries" in such a manner so as to accept the claims of Florida and Texas.[40]

Another definition used in this act was later to cause some confusion as to the exact nature of the claim embodied therein. The term "natural resources" was said in the act to include "oil, gas, and all other minerals, and fish, shrimp, oysters, clams, crabs, lobsters, sponges, kelp and other marine animal and plant life . . . "[41] The defintions contained in a statute apply to the wording throughout unless there are stipulations to the contrary. In connection with this definition of "natural resources," note should be taken of section nine of the act, which reaffirms the rights of the United States government to the natural resources of the continental shelf seaward of the area given to the states. At the 1958 Geneva Conference on the Law of the Sea, a representative of Mexico raised the point that under this legislation, the United States had included "fish, shrimps, lobster, crabs, sponges, and pearl oysters among the natural resources of the Continental shelf."[42] Majorie Whiteman, in replying to this statement for the United States, observed that the Submerged Lands Act related solely to the American territorial sea and to the land beneath it. According to Whiteman, the act bore no relation to the continental shelf beyond the limit of territorial waters.[43] In fact, section nine of the act does deal with the shelf beyond territorial waters to the extent that it notes that the natural resources in that area appertain to the United States. The thrust of the Whiteman comment is that the term "natural resources" as defined in this act was intended by American legislators to apply only to the submerged lands inside United States territorial waters; the implication is that "natural resources" means one thing with respect to the "inner" continental shelf and something quite different with respect to the "outer" continental shelf. An official long active in ocean affairs has suggested that this was, indeed, the intention of Congress in the Submerged Lands Act.[44] In view of this, the act may be seen as an example of a poorly drafted piece of legislation.

The Outer Continental Shelf Lands Act

Having dealt with offshore land beneath territorial waters, congressional attention now turned to the "outer" continental shelf. Accordingly, after due consideration, the Congress passed a bill which when signed by President Eisenhower became the Outer Continental Shelf Lands Act of 1953.[45] In its report on this act, the Senate Interior Committee noted that the purpose of the legislation "is to assert the exclusive jurisdiction and control of the Federal Government over the seabed and subsoil of the Outer Continental Shelf, and to provide for the development of its vast mineral resources."[46]

In the hearings before the Senate Interior Committee, Senator Daniel stated that the claim embodied in the Outer Continental Shelf Lands Act should be one of jurisdiction to the seabed itself rather than to the natural resources of the seabed as stated in the Truman Proclamation.[47] Questioning Jack Tate, the Deputy Legal Adviser of the State Department, Daniel asked if there were anything wrong with such a position. Tate's reply was that the senator's stance was entirely unnecessary. The point made by Tate was that what was desired was control and jurisdiction over the resources of the continental shelf. He was suggesting, as the Legal Adviser's Office had done all along, that it would be detrimental to American interests to claim more than was absolutely necessary. In addition, Tate noted the desirability of staying within the framework of the 1945 presidential proclamation, which had evoked no protest. A change in the terminology employed by the United States might engender confusion. Tate did admit, however, that the United States had not protested the claims of other states that "annexed" their shelf areas, provided that they were not making claims to the overlying high seas.[48] The misgivings which the State Department had on this matter were not shared by the Interior Committee and the final wording of the act does refer to the seabed itself.

A major concern that was shared by the Interior Committee and the State Department was that any claim to the continental shelf should not affect the international status of the high seas above the shelf. Congress did recognize that international law limited the scope of an offshore claim in this regard. Texas Senator Daniel noted that "beyond our territorial waters or historic boundaries, the overlying waters on top of the Continental Shelf belong to the family of nations and the United States cannot have exclusive rights there."[49] Senator Cordon also voiced strong support for the need to emphasize the distinction in legal status between the shelf and the waters above it.[50]

This differentiation was supported by the American fishing industry. Speaking on behalf of the American Tunaboat Association and other fishery organizations, Wilbert Chapman asserted that if one state could unilaterally change its territorial limits in the high seas even slightly, then another state had the equal right to extend its boundaries out into the high seas by as much as it might consider necessary for its own welfare. In light of this consideration, Chapman feared the consequences of an American abandonment of the three-mile limit and the subsequent damage that would result to the American fishing industry.[51]

Congress heeded the warnings of members of the Interior Committee, State Department, and the fishing industry and provided in the Outer Continental Shelf Lands Act a clause stating "This Act shall be construed in such manner that the character as high seas of the waters above the outer Continental Shelf and the right to navigation and fishing therein shall not be affected."[52]

In line with the desire to leave the seas above the continental shelf unaffected by this legislation, the Interior Committee accepted the advice of Deputy Legal Adviser Tate not to use the term "sovereignty" in relation to the shelf. In place of that word he urged the adoption of "jurisdiction and control" since this expression would make it clear that the United States continued to support the principle of freedom of the seas. Tate did acknowledge, however, that "sovereignty" and "jurisdiction and control" were tantamount to the same thing. Senator Daniel then stated that as long as this were the case, he would accept Tate's wording.[53]

The extent of the area to which Congress was asserting a claim on behalf of the United States corresponded to what was understood to be the geographical continental shelf. According to the Senate Interior Committee:

> The Continental Shelf is defined as the extension of the land mass of the continents out under the waters of the oceans to the point where the continental slope leading to the true ocean bottom begins. This point is generally regarded as a depth of approximately 100 fathoms, or 600 feet, more or less.[54]

The depth criterion as stated above is similar to that associated with the Truman Proclamation. Further, it was not the intention of Congress to include the continental slope within its definition of the continental shelf; in the congressional view, the shelf ended where the slope began. Since this act was seen as an implementation of the Truman Proclamation, such a demarcation of the shelf is understandable. We should note, however, the additional consideration that available technology severely limited the range of possible operations. It was felt that a claim to the 600 foot isobath would be quite sufficient for all practical purposes in the forseeable future.

The Outer Continental Shelf Lands Act provided statutory authority for the granting of leases in the outer shelf area and gave power to the Secretary of the Interior to administer the leasing and to prescribe such rules and regulations as were necessary.[55]

Developments in the Inter-American System

The sharp differences between the United States and a number of Latin American states as to the nature and extent of claims that could be made to offshore areas revealed themselves in conferences held under the auspices of the Organization of American States. At its first meeting in Rio de Janeiro in 1950 the Inter-American Council of Jurists requested the Inter-American Juridical Committee to examine the matter of territorial waters and related questions. The Committee did so and in 1952 presented a draft treaty consisting of five articles. Most significantly, the draft articles asserted that existing international law recognized the exclusive sovereignty of the coastal state over the soil, subsoil, waters, and airspace above the continental shelf. States that became signatories to this draft convention further recognized the right of each state to claim an offshore zone for exclusive economic exploitation to a distance of 200 miles from the coast.[56]

This draft was extremely controversial and of the committee's seven members only a bare majority endorsed it.[57] Charles Owen, the American and one of the dissenters, noted that the signatories to this document were expected to recognize rights that contemporary international law simply did not support. Finding the draft totally unacceptable, he successfully endeavored to produce a single dissenting opinion signed by the three dissatisfied committee members. In his view a unified dissent would have greater impact than would separate dissents, each having a different slant. The dissent that was produced claimed that the committee had not followed its mandate; it said that the committee should have prepared a report rather than a draft convention and that it could not be seriously maintained that international law recognized the principles contained therein.[58] No doubt after the exertion of considerable pressure by the United States, this draft treaty was returned to the committee by the second meeting of the Council of Jurists so that it could be studied further.[59]

At the tenth Inter-American Conference held in Caracas in 1954, it was resolved that a conference be called to study the matter of off-shore jurisdiction and rights. The Council of the Organization of American States also requested pertinent bodies to render the necessary cooperation in the preparatory work for this conference.[60]

In accordance with this request, the Inter-American Council of Jurists at its third meeting in Mexico City in January of 1956 adopted a resolution that was said to be a preparatory study for the specialized conference that was soon to meet in Ciudad Trujillo. The resolution stated in Part A that:

1. The distance of three miles as the limit of territorial waters is insufficient, and does not constitute a general rule of international law. Therefore, the enlargement of the zone of the sea traditionally called 'territorial waters' is justifiable.
2. Each state is competent to establish its territorial waters within reasonable limits, taking into account geographical, geological, and biological factors as well as economic needs of its population, and its security and defense.[61]

The United States was very strongly opposed to this resolution seeing in it a challenge to the most basic American conceptions of the law of the sea; nevertheless, it was outvoted by a margin of 15-1. For the record, the United States made known its very serious doubts about the resolution. The point was made that the council did not have available to it preparatory studies that were indispensible before decisions could be made. Detailed discussion of the council's resolution was lacking. Much of the resolution was said to be contrary to international law. Only the rights of adjacent coastal states were considered in relation to the conservation and utilization of marine resources; the need for international cooperation was totally neglected. Lastly, the United States claimed that the resolution was "clearly designed to serve political purposes and therefore exceeds the competence of the Council of Jurists as a technical-juridical body."[62]

The United States was bitterly disappointed with the results of the Mexico City Conference. Marjorie Whiteman, Assistant Legal Adviser of the State Department, noted that at this conference Chile, Ecuador, and Peru remained undaunted in their assertions of sovereignty and jurisdiction over large areas of the high seas and in fact had spread their influence. Attacking the argument that these states were merely trying to protect and conserve offshore food resources from destruction by states with more advanced fishing fleets, she claimed that they were motivated instead by a desire to raise revenue from the use of the high seas. Whiteman reiterated that the Mexico City resolution was "made in the absence of the submission of data, scientific or legal, with respect to its alleged basis in law or fact." As an attempt to further deprecate the significance of the resolution, she emphasized that it was merely a "preparatory study" for the coming Ciudad Trujillo Conference.[63]

Because of its dissatisfaction with the earlier Inter-American conferences, the United States wanted to make certain that the results of the Ciudad Trujillo Conference would be more to its liking. Toward this end the United States delegation to this conference provided a number of technical papers to help guide the conference representatives.[64] The United States was thus attempting to provide what it felt to be a more factual basis upon which to make judgments. As a result of past disappointments the State Department carefully limited its objectives, desiring only that the Ciudad Trujillo Conference examine and discuss the scientific and legal aspects of offshore problems without recommending conclusions as to the seaward limits of national sovereignty.[65] The State Department was well aware of the pressure of certain Latin American states to seek a 200 mile zone of offshore jurisdiction and apparently felt that this matter was appropriately a problem that should be dealt with by the United Nations. The United States could expect greater sympathy in United Nations bodies, the membership of which included a number of European states with views similar to those of the United States, than it could in an inter-American arena where the Latin American states had the strength of numbers on their side.[66]

The resulting resolution of the Ciudad Trujillo Conference was much more to the liking of the United States than were the resolutions of earlier conferences. Those items which could be agreed upon were agreed upon; on those points where there was a sharp disagreement, the disagreement was recognized as such and the minority view was not ignored the way it had been at Mexico City. The following conclusions were submitted to the American states for consideration:

1. The seabed and subsoil of the continental shelf, continental and insular terrace, or other submarine areas, adjacent to the coastal state outside the area of the territorial sea, and to a depth of 200 meters or, beyond that limit, to where the depth of the superjacent waters admits of the exploitation of the natural resources of the seabed and subsoil appertain exclusively to that state and are subject to its jurisdiction and control.

2. Agreement does not exist among the states here represented with respect to the juridical regime of the waters which cover the

said submarine areas, nor with respect to the problem of whether certain living resources belong to the seabed or to the superjacent waters.[67]

With regard to the first point above, three things should be noted. First, mention is made not only of the continental shelf itself but also of the continental and insular terrace, which includes the continental slope, and other submarine areas. There is here a departure from the geographical and geological notion of the continental shelf. That the United States delegation favored this resolution with this wording seems to reflect a concession made to the Latin American states. In a position paper prepared for the United States delegation to the conference, it was pointed out that very little was known about the continental slope and that technology did not then, nor would it for many years, allow exploitation there.[68] When, during the conference, the representative of Panama first called for the inclusion of the slope within the area to be brought under national jurisdiction and control, the American delegate noted that this might be too great an extension of national authority.[69] Nevertheless, the United States accepted the wording of the Ciudad Trujillo Resolution. According to an adviser on the American delegation at that conference, delegation concern was primarily directed toward problems such as territorial waters and conservation zones. The question of the continental shelf did not seem to generate controversy.[70] Since in 1956 the deepest water in which any exploitation of oil or gas had been attempted was in the 150-200 foot range, the continental slope was probably not thought to be of any great immediate importance and was thus an area in which the United States could compromise.

Second, it is interesting to note that the term "jurisdiction and control" is employed rather than the term "sovereignty" as used in the 1952 draft articles of the Inter-American Juridical Committee. The Ciudad Trujillo terminology thus parallels that of the Truman Proclamation and was clearly favored by the State Department.

Third, at Ciudad Trujillo the "exploitability" clause reappeared after having been dropped by the International Law Commission in 1953.

The second point of the Ciudad Trujillo Resolution noted that agreement did not exist with respect to the regime of the waters above the submarine areas or with respect to the problem of what living resources belong to the continental shelf. The United States was willing to recognize that some living resources were continental shelf resources. A study prepared for the American delegation suggested that those living resources attached to the bottom in the stage of life in which they were harvested might be declared continental shelf resources. The general nature of this recommendation indicates that the United States had not made a determination as to exactly what species were to be so designated.

In light of the Mexico City Resolution, voted by a 15-1 majority, it is interesting to consider how the much more moderate Ciudad Trujillo Resolution was approved by the same group of states. Why did the Latin American states not reassert the views thay had so recently put forward? Surely it was not the American technical studies, despite their high quality, which kept the previously vocal and determined Latin American states from reaffirming as legitimate the legal principles they proclaimed at Mexico City. In fact, we find that the United States brought very intense diplomatic pressure to bear on these states during the short interval between the Mexico City and Ciudad Trujillo Conferences. The Latin American states were explicitly informed that the United States "would not tolerate a repetition of the Mexico City meeting and that it would have to reconsider its membership in the Organization of American States unless positions were modified."[71] Positions were not necessarily modified but a resolution acceptable to the United States was produced. Within the American government, it was noted that "the lesson to be learned is that it was diplomatic pressure and not biology which won the battle."[72]

The profound satisfaction with the Ciudad Trujillo Resolution was underlined by a United States delegation statement at the conclusion of the conference, which noted that the resolution "constitutes the latest and most authoritative expression of the Organization of American States on the subjects discussed therein."[73] In making this statement the United States was clearly attempting to undermine the significance of earlier inter-American resolutions and draft treaties which contradicted this later resolution and the views of the United States.

DEVELOPMENTS IN THE INTERNATIONAL LAW COMMISSION

At its first session in 1949, the International Law Commission composed a list of international law topics for codification and among them was that of the regime of the high seas. Early in 1950, the ILC circulated the replies it had received to a questionnaire sent to states belonging to the United Nations. The United States memorandum on the high seas made two major points. First, it noted that the United States from the beginning of its history had insisted upon the three-mile rule of territorial waters. In some detail it demonstrated the historical veracity of this fact.[74] Second, the United States strongly emphasized that the 1945 Truman Proclamations had no effect whatsoever on free navigation in the waters over the shelf or in the fishery conservation zones.[75]

At its third session the commission, having considered the replies to the aforementioned questionnaire, drew up several draft articles on the continental shelf and related subjects. Article one defined the continental shelf as the offshore seabed and subsoil of the "submarine areas" contiguous to the coasts and extending beyond territorial waters to where "the depth of the superjacent waters admits of the exploitation of the natural resources of the sea-bed and subsoil."[76] In its commentary on this article, the ILC noted that the continental shelf as described departed from the geological concept of that term.[77] According to the ILC, the departure was necessary since there were offshore areas in shallow waters that might not qualify in a geological sense as "continental shelf" but might, nevertheless, allow exploitation of the seabed as if it were a continental shelf.[78] The commission did decide, however, to retain the term "continental shelf" because it was in current usage and because the expression "submarine areas" used alone would not clearly define the nature of the offshore areas in question.[79]

In the draft of article one we find the first official appearance of the "exploitability clause." In adopting this clause the commission explicitly rejected a fixed depth limit, such as 200 meters, because it felt that such a limit would have the "disadvantages of instability." It was believed that as technology developed a given limit would have to be constantly revised; thus the exploitability clause was favored.[80] In light of contemporary conflicts it seems somewhat ironic that a fixed limit was rejected because of its supposed potential for problems.

Article two of the ILC draft stated that the continental shelf "is subject to the exercise by the coastal State of control and jurisdiction for the purposes of exploring it and exploiting its natural resources." The ILC specifically rejected any reference to "sovereignty," noting in its commentary that, "As control and jurisdiction by the coastal State would be exclusively for exploration and exploitation purposes, they cannot be placed on the same footing as the general powers exercised by a State over its territory and its territorial waters."[81] Further supporting the basic view of the United States, draft articles three and four asserted that coastal state jurisdiction and control over the shelf does not affect the legal status of the waters or air space above.

In addition, the ILC also expressed the opinion, in line with that of the United States government, that the problems of fishing and the conservation of the resources of the sea should be dealt with separately from the issue of the continental shelf.[82] At this time the commission felt that the separation should extend even to sedentary fisheries.

Having drafted these articles, the commission invited comments from governments. In formulating a reply, the State Department called upon the Department of the Interior for an expression of its views. The Interior Department showed strong enthusiasm for the draft articles. Interior Secretary Chapman was of the opinion that it was in the American national interest to encourage the full use of the resources of the shelf areas off American coasts. He asserted that the proposed definition of the continental shelf served such an interest since it eliminated "such limiting factors as geological concepts, depth of superjacent waters, and distance from the coast."[83]

Again we see manifested the desire of the Interior Department to extend its domain, encouraging the growth of the continental shelf doctrine beyond the confining bounds of the geological continental shelf. Chapman claimed that the ILC perception of the shelf was fully consistent with the Truman Proclamation since neither the ILC draft nor the proclamation contained any geological, water depth, or distance from the coast concept with respect to the extent of the shelf.[84] In light of the White House press release of September 28, 1945 and the preambular portion of the Truman Proclamation, it may well be maintained that there was a geological and a water depth concept inherent in the proclamation.

For its part, the State Department must have been pleased with the result of the work of the ILC. Concerned as it was with the legal status of the epicontinental waters and air space and the claims which some Latin American states had made thereto, the commission's draft was reassuring and supportive of its position. It is thus not surprising that the United States government expressed its general agreement with the principles contained in the ILC draft.[85]

At its fifth session in 1953, the ILC reexamined the draft articles it had adopted in 1951 in light of governmental comments. As a result of its deliberations, the commission adopted eight draft articles on the continental shelf.[86] The first article of this new draft provided a definite outer limit of 200 meters for the continental shelf; the exploitability clause of the earlier draft had been removed. In explaining this change the ILC noted in its commentary that the exploitability clause lacked clarity and could lead to disputes. The 200-meter limit would be sufficient for all practical purposes and had further been adopted because "it is at that depth that the continental shelf, in the geographical sense, generally comes to an end. It is there that the continental slope begins and falls steeply to a great depth."[87] This observation seemed to indicate a movement back to the geological and geographical concept of the shelf but the ILC commentary goes on to assert that "the Commission in no way holds that the existence of the continental shelf in its geographical configuration as generally understood is essential for the exercise of the rights of the coastal State as defined in these articles."[88]

Article two of the new draft also contained changes. It now read "The Coastal State exercises over the continental shelf sovereign rights for the purpose of exploring and exploiting its natural resources." Whereas in the previous draft the shelf had been subject to the control and jurisdiction of the coastal state, it was now said that the coastal state exercises "sovereign rights" for certain limited purposes. In substituting this wording, the commission wished to make it quite clear that the coastal state possessed all rights necessary for and connected with the exploration and exploitation of the natural resources of the continental shelf. At the same time, by using the words "sovereign rights" rather than the term "sovereignty," the ILC sought to avoid language that might be misinterpreted so as to interfere with the principle of the full freedom of the superjacent sea and the air space above it.[89]

At this same session the ILC changed its position on yet another point. Observing that it had in the past considered only mineral resources of the shelf as natural resources, the commission broadened its perspective to include sedentary fisheries. Thus, for the first time, living resources were considered by the commission as natural resources of the continental shelf.[90]

In 1956 at its eighth session, the ILC adopted a draft treaty of 73 articles on the law of the sea.[91] At this session the commission again altered its position on the definition of the continental shelf, once more inserting an exploitability clause. In explaining the change back to the exploitability clause, the ILC noted that it was following the lead of the Ciudad Trujillo Conference.[92]

Having drafted this convention on the law of the sea, the ILC recommended to the General Assembly that the Assembly call an international conference to examine the matters dealt with by the commission. At its eleventh session the General Assembly passed such a resolution.[93]

In the period between 1945 and 1958 we find a growing interest, both domestically and internationally, with respect to offshore areas. Domestically, the matter of federal-state jurisdiction over portions of the continental shelf was dealt with by the courts and by Congress. Congress also provided for commercial utilization of the outer continental shelf. At this point we see a limited but developing interplay of forces involving the Interior and State Departments and fishery and oil interests. Internationally, the United States found itself attempting to maintain the careful balance it sought to create with the Truman Proclamation, a balance of national access to mineral resources of the shelf but denial of exclusive rights in the epicontinental seas and in the air space above the shelf. Toward the end of achieving acceptance of its legal position, the United States found it necessary vigorously to protest and combat the more extensive claims of Latin American states. Through the use of diplomatic pressure, the United States was able to achieve acceptable results at Ciudad Trujillo, but it remains clear that the Latin American states were not converted to the views held by the United States. Finally, the United States contributed to the shaping of the ILC draft treaty on the law of the sea.

NOTES

1. This information is contained in a letter to the author.
2. United States Senate, Committee on Interior and Insular Affairs, "Outer Continental Shelf," 83rd Congress, 1st session, p. 573.
3. White House press release, September 28, 1945, "Proclamations Concerning United States Jurisdiction Over Natural Resources in Coastal Areas and the High Seas."
4. Wilbert Chapman, "United States Policy on High Seas Fisheries," United States Department of State Bulletin 20 (1949): 20. See also Douglas Johnston, The International Law of Fisheries (New Haven: Yale University Press, 1965), pp. 333-334.
5. United States Senate, Committee on Interior and Insular Affairs, "Outer Continental Shelf," op. cit., p. 601.
6. White House press release of September 28, 1945.
7. Executive Order 9633 of September 28, 1945.
8. Annual Report of the Secretary of the Interior, 1945 (Washington, D.C.: Government Printing Office, 1945), p. x. In this report Ickes wrote that "Approximately described, the Continental Shelf is all of the ocean floor around the United States and its Territories that is covered by no more than 600 feet of water."
9. Harold Ickes, "Underwater Wealth," Collier's, February 23, 1946, p. 20.
10. Speech of John Stevenson before the Offshore Technology Society, Houston, Texas, April 19, 1971, p. 11. See also the speech of Defense Department spokesman Leigh Ratiner before this conference, especially p. 10.
11. The text of this declaration appears in Naval War College International Law Documents, 1948-1949 (Washington, D.C.: Government Printing Office, 1950), pp. 185-186.
12. Ibid.
13. Ibid.
14. Letter of the Acting Secretary of State to the Mexican Ambassador, dated January 24, 1946. Foreign Relations of the United States, 1946 11: 1054-1058.
15. Ibid.

16. The text of this document appears in Naval War College, International Law Documents, op. cit., pp. 187-188.

17. Ibid.

18. The text of this note of protest is reproduced in A/CN. 4/19, pp. 115-116.

19. The text of the Chilean Presidential Proclamation of June 25, 1947 appears in Naval War College, International Law Documents, op. cit., pp. 188-190.

20. The text of the Peruvian Presidential Decree of August 1, 1947 appears in ibid., pp. 190-192.

21. The text of the Costa Rica Decree Law of July 27, 1948 appears in ibid., pp. 193-195.

22. These notes of protest are found in A/CN. 4/19, pp. 113-115. Both are dated July 2, 1948.

23. Naval War College, International Law Situation and Documents, 1956, p. 452.

24. The text of this article appears in ibid., p. 460.

25. U.S. Department of State Bulletin 24 (1951):24.

26. Agreement Between Chile, Ecuador and Peru signed at the First Conference on the Exploitation and Conservation of the Maritime Resources of the South Pacific, Santiago, 18 August 1952. Text in Naval War College, International Law Situations and Documents, 1956, pp. 265-267.

27. Agreement Between Chile, Ecuador and Peru signed at the Second Conference on the Exploitation and Conservation of the Maritime Resources of the South Pacific, Lima, 4 December 1954. Ibid., pp. 275-282.

28. Herman Phleger, "Recent Developments Affecting the Regime of the High Seas," U.S. Department of State Bulletin 32 (1955): 234.

29. Ernest R. Bartley, The Tidelands Oil Controversy (Austin: University of Texas Press, 1953), pp. 122-143. See also R. Krueger, "State Tidelands Leasing in California," U.C.L.A. Law Review 5 (1958): 427-489.

30. Veto message of President Truman of May 29, 1952, Senate Document 139, 82nd Congress, 2d session, veto of Senate Joint Resolution 20. The message notes that Secretary Ickes, after studying the matter, concluded that the federal government had interests in the tidelands. See also Bartley, op. cit., pp. 135-140.

31. U.S. 19. Rulings similar to that in the California case were made by the Supreme Court with regard to the claims of Texas and Louisiana to the offshore lands adjacent to their coasts in United States v. Texas 339 U.S. 707 and United States v. Louisiana 339 U.S. 699.

32. Defense Secretary Forrestal, Attorney-General Clark, Interior Secretary Krug, "Explanatory Statement Concerning a Proposed Bill to Promote the Development and Conservation of Certain Resources in the Submerged Coastal Lands Adjacent to the Shores of the United States," dated February 2, 1948. In the Congressional Record, February 18, 1948, pp. 1356-1359.

33. United States Senate, Interior Committee, "Submerged Lands," 83rd Congress, 1st session, pp. 974-1049.

34. Public Law 31, 83rd Congress, 1st session.

35. Senate Interior Committee, "Submerged Lands," op. cit, pp. 124-125.

36. The letter is dated December 30, 1949 and appears in ibid., pp. 321-323.

37. See the testimony of John J. Real, Manager and Attorney for the Fishermen's Cooperative Association of San Pedro, California in ibid., pp. 295-325 and the letter of Congressman John F. Kennedy to the Senate Interior Committee, dated March 5, 1953, pp. 1177-1178.

38. Letter of Assistant Secretary of State Thruston Morton to the Senate Interior Committee, dated March 4, 1953, ibid., pp. 27-28.

39. Ibid., pp. 512, 527-528, 535-536.

40. Submerged Lands Act, Public Law 31, 83rd Congress, 1st session, section 2 (b).

41. Ibid., section 2 (e).

42. See the comments of Gomez Robledo of Mexico, A/Conf. 13/42, p. 15.

43. A/Conf. 13/42, p. 20.

44. This information derived from an interview in Washington, D.C.

45. Public Law 212, 83rd Congress, 1st session.

46. Senate Report 411, 83rd Congress, 1st session.

47. Senate Interior Committee, "The Outer Continental Shelf," 83rd Congress, 1st session, p. 587.

48. Ibid., pp. 587-592. Tate indicated that Congress had the constitutional authority to exercise full jurisdiction and control over the seabed and subsoil but questioned the need for this existence of authority. He claimed that for all practical purposes the American claim of the Truman Proclamation was just as inclusive as the claims of those states that dealt directly with the seabed and subsoil.

49. Ibid., p. 160.

50. Ibid., p. 9.

51. Ibid., p. 380. See also the testimony of W. L. Hardee of the Texas Shrimp Association on this point, pp. 243-245.

52. Public Law 212, 83rd Congress, 1st session, section 3 (b).

53. Senate Interior Committee, "The Outer Continental Shelf," op. cit., p. 589.

54. Senate Report 411, 83rd Congress, 1st session, p. 4.

55. Public Law 212, 83rd Congress, 1st session, sections 5 and 6.

56. The text of this draft treaty appears in Senate Interior Committee, "The Outer Continental Shelf," op. cit., p. 722.

57. Committee members from Mexico, Peru, Chile, and Argentina voted affirmatively while those from Brazil, Colombia, and the United States voted negatively.

58. Letter to the Department of State from George C. Owen, United States member, Inter-American Juridical Committee, dated August 25, 1952 found in the files of the Solicitor, United States Department of the Interior. Owen noted that due to the need for compromise to secure a unified dissent "some irrelevant hairsplitting remains" in the text of the dissent. The text of the dissent appears in Senate Interior Committee, "The Outer Continental Shelf," op. cit., pp. 724-727.

59. Resolution XIX, "Territorial Waters and Related Questions" of the Second Meeting of the Inter-American Council of Jurists appears in *Inter-American Juridical Yearbook, 1952-1954* (Buenos Aires, 1952), pp. 221-222.

60. Resolution LXXXIV of the Caracas Conference, Document 2, *Background Material on the Juridical Aspects of the Continental Shelf and Marine Waters*, pp. 25-26.

61. *Inter-American Juridical Yearbook*, 1955-1957, pp. 226-228.

62. The text of the United States statement appears in ibid., p. 239.

63. Marjorie Whiteman, "The Territorial Sea," *Proceedings of the American Society of International Law* 50 (1956): 127, 130-132. See also Richard Young, "Pan American Discussion of Offshore Claims," *American Journal of International Law* 50 (1956): 909-916.

64. See, for example, the papers prepared by O. E. Sette of the United States Fish and Wildlife Service, "Living Resources of the Continental Shelf and Their Utilization and Conservation," and E. M. Thomasson of the United States Geological Survey, "Technical Problems of Mineral Development on the Continental Shelves." Organization of American States, Official Records of the Ciudad Trujillo Conference, Document 36 and 28 respectively.

65. Memorandum of Ralph L. Miller, Chief of the Fuels Branch to the Director of the Geological Survey, Interior Department dated December 23, 1955. This memorandum concerning a meeting with State Department officials on continental shelf matters was found in the files of the Bureau of Land Management of the Department of the Interior.

66. Ibid.

67. Final Act of the Inter-American Specialized Conference on "Conservation of Natural Resources: Continental Shelf and Marine Waters," Pan American Union, 1956.

68. Position Paper on Continental Shelves, no specific author identified, dated February 13, 1956, in the files of the Secretary of the Interior, pp. 7, 9.

69. Organization of American States, Official Records of the Ciudad Trujillo Conference, Document 20, March 18, 1956 and Document 76, March 23, 1956.

70. This information is contained in letters to the author.

71. See the letter of Stillman Wright, chief, Office of Foreign Activities, to the Commissioner of the Fish and Wildlife Service, dated August 19, 1957. In the files of the Fish and Wildlife Service, Federal Records Center, Suitland, Maryland, Access number 67A-1358 Box 80.

72. Ibid.

73. The American statement is contained in the Final Act of the Inter-American Specialized Conference on "Conservation of Natural Resources: Continental Shelf and Marine Waters," op. cit.

74. A/CN. 4/19, pp. 104-105.

75. Ibid., p. 111.

76. The text of the draft articles appear in A/1858, pp. 17-19.

77. Ibid., p. 17.

78. Ibid.

79. Ibid.

80. Ibid., p. 80.

81. Ibid.

82. Ibid.

83. Letter of Interior Secretary Chapman to Secretary of State Acheson, dated July 30, 1951, in the files of the Department of the Interior, National Archives, Washington, D.C. Box 2 208, pt. 29.

84. Ibid.

85. The comments of most of the responding governments are found in A/CN. 4/55, May 16, 1952; the comments of the United States arrived after this date and are contained in A/CN. 4/55/Add. 2, June 21, 1952.

86. The text of these articles appear in A/2456, ch. III.

87. Ibid., para. 64.

88. Ibid., para. 65.

89. Ibid.

90. Ibid., para. 70.

91. The United Nations General Assembly had requested in Resolution 899 (IX) of December 14, 1954 that the ILC systematically group together all the rules it had adopted concerning the high seas, the territorial seas, the continental shelf, the contiguous zone, and

the living resources of the sea. Articles 67 through 72 of this draft concern the continental shelf. The text of the whole draft together with the ILC commentary is found in A/3159.

92. A/3159, para. 4.

93. General Assembly Resolution 1105 (XI), February 21, 1957.

CHAPTER

4

THE GENEVA CONFERENCE ON THE LAW OF THE SEA

On February 24, 1958 the Geneva Conference on the Law of the Sea convened with representatives of 86 states in attendance. The representatives who gathered in Geneva found a considerable body of preparatory work available to them, including the International Law Commission's draft articles on the law of the sea which served as the working paper of the conference.

American policy on continental shelf matters at Geneva must be seen in the broad context of American concern with the shaping of the law of the sea generally. Arthur Dean, chairman of the American delegation, has pointed out that the first goal of the United States at this conference was the preservation of the territorial sea at three miles.[1] While this may have been the major overall policy objective of the United States delegation, it is a fact that different actors within the American delegation were concerned with varying aspects of the work of the conference. The representatives of the United States were drawn primarily from three executive departments: State, Interior, and Navy (Defense).[2] It is worthwhile to consider the departmental points of view of the work at Geneva.

NAVY CONSIDERATIONS

Certainly for the United States Navy, the matter of the extent of territorial waters was seen as the question of greatest importance. As we have observed, it has been the traditional position of the Navy that the greater the freedom of the seas, the more protected would be the security interests of the United States. Captain Leonard Hardy, a member of the United States delegation, has indicated that a Navy analysis conducted in this period concluded that the value to the United States of a wider territorial sea was insignificant as compared to the damage that would be done to naval mobility were an enlarged territorial sea accepted.[3]

A more extensive territorial sea would mean that American naval vessels and aircraft would have to restrict their operations off foreign coasts. This would undermine the potential for diplomatic deployment of naval strength since such deployment was felt to be most effective when naval units could be seen from the shore.[4] Of greater significance, an expanded territorial sea might well mean that a number of important international straits could be closed to vessels and aircraft of the United States Navy and Air Force.[5] Naval analysis at the time of the Geneva Conference indicated that with the existing three-mile limit of territorial waters, American naval forces could go almost anywhere they might conceivably wish to go without entering the territorial waters of any foreign state.[6] American naval officers clearly felt that the attempt of the Soviet Union to extend the limit of territorial waters to 12 miles was firmly based on the desire to reduce the effectiveness of the seapower of the western alliance states.[7]

Additionally, the Navy was very concerned with the large Soviet submarine fleet. As Arthur Dean has observed, the striking power of enemy submarines would greatly increase were the territorial seas of neutral states enlarged.[8] Any factor that made the submarine more effective automatically gave an advantage to the Soviet Union which, even in 1958, had the world's largest submarine fleet. Furthermore, the Soviet Union was less vulnerable than was the United States to the effects of conventional submarine warfare since its geographic position made it less dependent on waterborne commerce.

With regard to the continental shelf itself, Navy thought at this time was quite unsophisticated. The attention of that department was focused upon the matter of the limit of the territorial seas and the continental shelf was seen as an unrelated issue.[9] Captain Hardy has noted that at first the Navy paid little attention to the claims of the Latin American states because they involved extensions into wide sea areas that did not significantly limit naval operations. Navy apprehension developed, though, over the fear that such extensions would encourage wider territorial claims in areas where the Navy was directly concerned. As Hardy states, the United States Navy "did not particularly object to extensions of jurisdiction into high seas areas bordering the territorial sea if they did not interfere with navigation."[10] Navy feeling was that the continental shelf could be dealt with apart from the matter of territorial waters; continental shelf jurisdiction did not imply in any way extension of territorial waters. Consequently, the continental shelf was of subsidiary interest to the Navy in the general context of the law of the sea.

STATE DEPARTMENT CONSIDERATIONS

The State Department's perspective on the law of the sea reflected some elements different from the Navy's. The State Department was influenced, of course, by the security considerations of concern to the Navy. Additionally, State had to take into account matters not of direct interest to the Navy. Considerable friction between the United States and Latin American states had arisen due to the offshore claims of these latter states. When American vessels were seized for violating the national laws of these states, it was the State Department which was responsible for negotiating the release of such ships and their crews. It was the State Department that had to attempt to persuade the Latin American states that their actions were in contravention of existing international law and, thus, that their claims to broad areas of the high seas as exclusive fishing grounds should be abandoned.

American distant water fishermen pressed for recognition of only the narrowest band of water for exclusive fishing by the coastal state. Congress, subject to the political pressure of the fishing industry, was becoming increasingly concerned by the claims and actions of some of the Latin American states. In this context, Congress passed the Fisherman's Protective Act of 1954. The act calls upon the Secretary of State to take appropriate action on behalf of the vessel and crew of ships seized by foreign governments in waters claimed by foreign states, but not recognized by the United States as territorial waters, and provides for Treasury Department reimbursement for fines paid to secure the prompt release of such vessels and crews.[11]

The State Department was under continuous pressure from Congress to do something about the problems of American fishermen as they operated off the coasts of Latin American states. For its part, the fishing industry kept the pressure on State. At the time of the Geneva Conference the American fishing industry was much more united with respect to desired outcomes of international negotiations than it was in the 1960s. Dividing the American fishing industry into two parts, coastal fishing and distant water fishing, we find that basically only the distant water fishermen felt threatened during the 1950s. There was no substantial foreign fishing going on off the coasts of the United States at this time with the exception of salmon fishing. The pressure of the fishing industry at Geneva was directed toward securing American rights to fish for tuna, shrimp, and other fish off the coasts of foreign states. A government fishery expert who was at the Geneva Conference has asserted that fishery interests were the most powerful nongovernmental force influencing American decisions. The strong influence of the fishing industry, it was said, stemmed from its unified position, its good organization and leadership, and its political contacts in Washington.[12]

In 1958 it was feared that American distant water fishing would be vulnerable to attack along two lines. First, it was felt that the limit of territorial waters could be extended or that extensive fishery jurisdiction zones could be created. American fishermen wished to make certain that continental shelf jurisdiction did not overlap into the waters above the shelf and that exclusive fishing zones were effectively limited. Second, there was concern on the part of these fishermen that there could be a liberal classification of the living resources of the sea as "resources of the continental shelf," thus placing them within the exclusive domain of the coastal state. Briefly stated then, American fishermen wanted to keep the matter of fishery rights apart from that of the continental shelf.

The scientific community was also interested in developments pertaining to the continental shelf. In 1954 the International Council of Scientific Unions had expressed the fear that the ILC draft articles would lead to interference with scientific research in offshore areas. The ICSU expressed the hope that the articles would be amended so as to insure that basic oceanographic research could be conducted without having to face "vexatious obstruction."[13]

This position was shared and supported by the American scientific community. An official of the National Academy of Science indicated that the NAS Committee on Oceanography urged the American delegation at Geneva to support the right for anyone to carry out research on the waters above or on the seabed and subsoil of the continental shelf under the following conditions: 1) the research is openly published; 2) the adjacent state has the right to participate and observe the scientific investigation if it so desires.[14]

INTERIOR DEPARTMENT CONSIDERATIONS

While the Navy focused its attention at Geneva on the extent of territorial seas and the security implications thereof and State focused largely on these matters and fisheries problems, the representatives of the Interior Department were primarily concerned with the continental shelf. It was a point surely not lost upon Interior officials that whatever offshore land came under United States jurisdiction would come under the control of the Department of the Interior. More importantly, perhaps, there was no doubt that the American oil industry was greatly intrigued with the possibility of profitable exploitation opportunities that seemed to exist in offshore areas. This is seen in the fact that the first sale of leases under the 1953

Outer Continental Shelf Lands Act resulted in payments to the federal government of almost $130,000,000.[15] Interest was not limited to parcels of submerged land very close to the coast. As early as 1955, the Interior Department was able to lease land successfully as distant as 60 miles from shore.[16]

While reflecting the interest of the American oil industry in extending the area of offshore land under American jurisdiction, the Interior Department was also pleased with the financial benefits that the United States government derived from petroleum exploration there. Pointing to its successful lease sale, the Interior Department suggested that the leases open "the way for additional contributions to the Treasury of royalties and rents which conservatively are estimated to yield another half billion dollars over a period of years." It was further observed that "Development of the Shelf will give employment to many hundreds of men and women, will help to balance the Federal budget and will lighten the tax burden on citizens."[17] Seemingly, then, the Interior Department had great expectations for continental shelf development and the beneficial consequences thereof. A conclusion that might be drawn is that the greater the offshore area under American control, the more substantial the benefit to the United States Treasury and to the American public generally.

THE GENEVA CONFERENCE

Extent of the Continental Shelf

As it emerged from conference committee IV, and as found in article one of the 1958 Continental Shelf Convention, the article defining the continental shelf reads:

> For the purposes of these articles, the term "continental shelf" is used as referring (a) to the seabed and subsoil of the submarine areas adjacent to the coast but outside the area of the territorial sea, to a depth of 200 meters or, beyond that limit, to where the depth of the superjacent waters admits of the exploitation of the natural resources of the said areas; (b) to the seabed and subsoil of similar submarine areas adjacent to the coasts of islands.[18]

Though the United States had voted in favor of the 1956 Ciudad Trujillo Resolution, which included an exploitability clause, there still seemed to be some doubt within the American delegation at

Geneva about the desirability of such a clause. In a statement made by Marjorie Whiteman on behalf of the United States it was noted that the United States accepted the depth of 200 meters as an appropriate definition of the limit of the continental shelf; she expressed reservations, however, about the wisdom of adopting an exploitability clause.[19] Later in the conference she reflected a somewhat different United States position observing that, while she appreciated the desire on the part of some for precision, she doubted whether in an age of rapid scientific progress it would be wise to limit the area of exploitation in that way.[20]

Several factors led to the decision of the United States to favor exploitability. The Geneva Conference dealt with many aspects of the law of the sea. On the international level, compromise and bargaining had their effect. A concession made to one state in regard to fisheries might be necessary for that state to gain a vote or support in another area of greater interest to itself. The matter of the continental shelf was not immune to such activity. As seen, the Latin American states at Ciudad Trujillo favored the exploitability criterion and, in part, to gain support in other areas of conference concern, the United States accepted it. At the Geneva Conference also the Latin American states tended to support any formula which extended coastal state jurisdiction.[21]

William C. Herrington, who served on the United States delegation, has claimed that exploitability:

> Was a compromise concept. The coastal States who wished to extend their jurisdiction hoped this would provide them with a basis to continue to go out as they learned how to exploit the resources beyond that point [200-meter isobath]. The proponents of limited jurisdiction hoped this would be enough to stop the extension of jurisdiction.[22]

While attempting to gain some Latin American support by favoring exploitability, the United States delegation, at least in part, felt that the exploitability criterion had significant potential to limit coastal state jurisdiction. In the view of the American delegation the exercise of sovereign rights beyond the 200-meter water depth would have to await the development of the technology of exploitation which did not then exist.

On the domestic level, it is interesting to note that representatives of the Interior Department and the oil industry favored the exploitability clause because of the possibility of extension of coastal state jurisdiction presented thereby.[23] The petroleum industry seemed to have greater faith than did the diplomats in overcoming the technical problems of working in deep water. Thus it is seen that Herrington's comment cited above, while having validity in the interstate context he suggests, also seems to apply to the internal views of the American delegation.

The State Department, additionally, may have been somewhat favorably influenced toward the exploitability provision by a factor observed by Bernard Oxman. According to Oxman, some maritime states favored the limit of exploitability rather than a fixed limit to the continental shelf because of the fear that the 200-meter isobath, if universally accepted for one purpose, might then become a more credible fixed limit for other purposes as well, particularly, it was felt, for coastal state fisheries. From the point of view of maritime states with important naval, shipping, or noncoastal fishing interests, a vague exploitability test which applied to seabed areas likely to remain unexploitable for many years to come was a small price to pay for "holding the line" of coastal waters jurisdiction at three to twelve miles.[24]

Scope of Authority Over the Shelf

Article two of the Continental Shelf Convention reads in part: "The coastal State exercises over the continental shelf sovereign rights for the purpose of exploring and exploiting its natural resources." This wording, adopted in plenary session, is different from the wording adopted in committee at the Geneva Conference. In committee, an American amendment had been approved by a very narrow vote changing "sovereign rights" to "exclusive rights." The intention of this amendment was to safeguard the interests of both the United States Navy and American fishermen, who feared the use of terminology that might create rights in the waters above the continental shelf. Whiteman has noted that at the time this matter was being dealt with in committee, it was impossible to forecast the fate of the article of the International Law Commission draft treaty that protected the existing status of the waters above the shelf and that of the air space above these waters. In light of this fact, the United States found it desirable to "play it safe" by avoiding the use of the term "sovereignty" or even "sovereign rights" in defining the relation of the coastal state to the continental shelf.[25]

By the time the matter had come to the attention of the plenary session, it was evident that the legal status of the superjacent waters and air space was to be unaffected by the articles on the continental shelf. It was felt that Navy and fishing interests were protected by article three of the Continental Shelf Convention, which states that "The rights of the coastal State over the continental shelf do not affect the legal status of the superjacent waters as high seas, or that of the airspace above those waters." Since these important rights were thus protected, the United States, in an attempt to win increased support for the continental shelf articles, supported in plenary session a change in wording from "exclusive rights" to "sovereign rights."[26] Note that according to the wording of article two, the coastal state exercise sovereign rights for the limited purpose of natural resource exploration and exploitation.

Definition of Natural Resources

Article two, paragraph four of the Continental Shelf Convention provides a guideline as to what is to be considered as a "natural resource of the continental shelf." This provision reads:

> The natural resources referred to in these articles consist of the mineral and other non-living resources of the sea-bed and subsoil together with living organisms belonging to sedentary species, that is to say, organisms which, at the harvestable stage, either are immobile on or under the sea-bed or are unable to move except in constant physical contact with the sea-bed or subsoil.

The above wording was carefully considered and vigorously debated at the Geneva Conference. No problem arose in relation to the matter of mineral resources of the shelf, but substantial differences were in evidence in regard to living organisms. Italy, for example, expressed the view that the term "natural resources" should apply only to "inorganic natural resources"[27] while Iceland, on the other hand, claimed that the coastal state should have jurisdiction over free-swimming fish in the waters above the continental shelf.[28] A number of states favored the inclusion of, at least, "sedentary species" of living organisms as natural resources of the continental shelf.[29] A distinction should be recognized between "sedentary species" and "sedentary fisheries." Sedentary species refer to types of living organisms that have a natural and close relationship to the sea bottom. Sedentary fisheries refer to the fishing process that uses fixed gear attached to the sea bottom. Sedentary fisheries as such are treated in article 13 of the Convention on Fishing and Conservation of the Living Resources of the High Seas.[30]

The American position on living resources of the continental shelf was cautious. It was obvious that there was wide support at Geneva for the inclusion of some living species as shelf resources. According to the representative of India, if the term "natural resources" were not interpreted "reasonably," there might well be made wider claims affecting the character of the superjacent waters as high seas.[31] This sentiment was not lost upon the American delegation. Indeed, Wilbert Chapman has stated that the United States Navy used fish as a bargaining item with other states so as to safeguard its own interests in protecting free mobility.[32]

In the broad context of general fishery problems, the United States found itself at Geneva caught between those states that wished to secure maximum control by coastal states over fishery resources and states which believed their interests to lie in the direction of maintaining past freedoms with minimum restrictions. American fishermen, not successful in winning conference support for the principle of abstention,[33] tended to the latter point of view.

The American position attempted to take into account the preferences of the many states favoring the inclusion of some living resources as shelf resources and the opposing preferences of the American shrimp industry, which feared any such inclusion. Speaking on behalf of the United States, Marjorie Whiteman observed that the expression "natural resources" obviously included mineral resources since shelf mineral resources had the same origin as those found on land. Living resources, though, were basically products of the water and might be seen as belonging to the high seas. The best criterion for determining what living resources might be recognized as natural resources of the shelf appeared to be that of attachment to the seabed during the harvestable stage of life. Other criteria, it was suggested, would pose difficult problems of definition, might interfere with freedom of the high seas, and prevent the proper development and conservation of the resources concerned.[34]

It was well understood by members of the American delegation that providing a widely accepted and precise definition of living resources of the shelf would be difficult. The criterion of attachment to the seabed during the harvestable stage of life was by no means problem-free. In a preparatory study done within the United States government it was determined that:

> From a purely biological view it is difficult to distinguish between degrees of bottom dependence. Thus, any distinction, any attempt to separate the living forms into resources of the shelf and resources of the sea on the degree of bottom dependency would be arbitrary.[35]

William Herrington of the United States delegation at Geneva stated that he agreed in part with those who maintained that the term "mineral resources" should be used rather than "natural resources." He felt that the use of the former term would provide a positive line of separation between the resources of the shelf and those of the sea. On the other hand, he was well aware of the political facts of life and suggested that it would be wrong to ignore the point that in practice a number of states actually enjoyed exclusive historical rights in respect to some organisms living on their continental shelves.[36] Herrington's explanation of exclusive rights of certain states to a limited number of unspecified species of living resources of the shelf in terms of historical right, as opposed to a more general right, reflected the American desire to limit fishery resources, which were to appertain exclusively to the coastal state. Moreover, the argument of historical rights could be used by American shrimpers. The continued American attempt to limit coastal state jurisdiction over fisheries is further seen in the announcement that the United States opposed all attempts to include fish that occasionally had their habitat at the bottom of the sea or were bred there as creatures of the shelf.[37] The United States thus wished "living resources of the continental shelf" to be very narrowly construed. In the American view, for example, shrimp did not qualify as such a resource.[38]

Scientific Research

The Continental Shelf Convention has been strongly criticized by the scientific community. One leading oceanographer has termed the convention a disaster for the principle of freedom of scientific research.[39] The United States delegation at Geneva must share some of the blame for this result despite the American assertion that the United States had supported the principle of freedom of scientific research in offshore areas.[40]

Article 71 of the International Law Commission draft, which provided the basis for article five of the Continental Shelf Convention, reads, in part: "The exploration of the continental shelf and the exploration of its natural resources must not result in any unjustifiable interference with navigation, fishing or the conservation of the living resources of the sea." At Geneva, the representative of Denmark suggested an amendment, which was approved, adding to the end of this paragraph the words "nor in any interference with fundamental oceanographic research carried out with the intention of open publication."[41] Though the Danish spokesman noted that his proposal referred to "fundamental oceanographic research only," the United States abstained despite its alleged support for the freedom of scientific research.[42]

The most damaging provision in the Continental Shelf Convention from the point of view of the scientific community is found in article 5 (8). It reads:

> The consent of the coastal State shall be obtained in respect of any research concerning the continental shelf and undertaken there. Nevertheless, the coastal State shall not normally withhold its consent if the request is submitted by a qualified institution with a view to purely scientific research into the physical or biological characteristics of the continental shelf, subject to the proviso that the coastal State shall have the right, if it so desires, to participate or to be represented in the research, and that in any event the results shall be published.

Assessing the impact of article 5 (8) in retrospect, William L. Sullivan, Jr. of the State Department has observed that the United States "made a colossal blunder in leading scientific research under this restriction. . ."[43] Wilbert Chapman asserts that at the conference it was the United States government that insisted upon the restrictive provisions of the Continental Shelf Convention in regard to scientific research. Attempts within the American delegation to mitigate such provisions met adamant opposition and failed.[44]

Restrictions on the freedom of scientific research were clearly favored by the Interior Department. In a letter from D. W. Bronk, President of the National Academy of Sciences, to the Department of State soon after the adoption of the convention, it was noted that the International Council of Scientific Unions, of which the Academy is a member, had adopted a resolution which called upon national members of the ICSU to make a request to their respective governments. Governments were to be asked to signify at the time they ratified the Continental Shelf Convention that they would grant general permission for scientific research in their continental shelf areas to any vessel whose program had been approved by the ICSU. The intent of this proposal was to provide assistance to governments in identifying legitimate scientific research and to eliminate delays that might jeopardize such research. Accordingly, the National Academy's Committee on Oceanography urged the State Department to consider this proposal favorably.[45]

In connection with the Bronk letter, the Legal Adviser's Office of the State Department requested the views of the Interior Department. Interior's reply was negative. According to that department it was necessary for the United States government to retain for itself the power to prohibit activities of foreign nationals in relation to the continental shelf areas adjacent to American coasts. It was further noted that under section 11 of the Outer Continental Shelf Lands Act the Interior Department was given responsibility for the authorization of research programs in such areas. Adoption of the proposal of the National Academy of Sciences, it was said, would delegate this power to an international body and deprive the United States government of any real opportunity to influence such decisions.[46] Consequently, Interior opposed the NAS suggestion; in large part this opposition stemmed from the fear of Interior that its control over American continental shelf areas would be lessened.

Professor Milner Schaefer has noted that subsequent to the nonacceptance of the ideas expressed in the Bronk letter, the Oceanography Committee of the National Academy of Sciences further investigated this matter with officials of the Departments of Defense, Navy, Interior, Commerce, and State. As a result of these consultations the National Academy of Sciences concluded that it would be impossible to get the cooperation of the United States government for the ICSU proposal.[47]

Military Uses

At the Geneva Conference, the Bulgarian delegation proposed to add a new paragraph to what became article five of the Continental Shelf Convention. The suggested addition read: "The coastal State shall not use the continental shelf for the purpose of building military bases or installations."[48] In response to questions concerning this proposed amendment, the representative of Bulgaria observed that this wording would not allow military bases or installations on the continental shelf for either offensive or defensive purposes.[49]

The Bulgarian initiative may well have been explicitly designed to cause embarrassment to the United States, which in 1955 had undertaken a program of installing Texas Tower radar stations on the wide continental shelf off the east coast of the United States.[50] These towers were designed to lessen the possibility of a surprise attack on the United States. Hanson Baldwin had pointed out at the time that radar effectiveness was then limited to the line of sight; about 100 to 150 miles was said to be the extreme range of shore based stations.[51] Thus at this period the Texas Towers may have been of some importance to American defense planners. By the middle of 1956 it was publicly known that a Texas Tower radar station had been operating for a full year 110 miles east of Cape Cod.[52]

The placing of these towers far beyond the limit of American territorial waters raised some question as to the legal basis for such an operation. According to the United States Air Force, which was responsible for the placement and operation of these towers, the legal basis for the construction of these facilities was considered in depth by various governmental agencies. In the opinion of the Air Force the legal justification for the towers was based upon the concept of reasonable self-defense uses of the high seas. As evidence of the reasonableness of its efforts, it was pointed out that the Air Force coordinated the placement of these towers with maritime interests to avoid interference with fishing, shipping, and other lawful uses of the high seas. According to the Air Force, the placement of the radar towers did not relate to the regime of the continental shelf but rather to that of the high seas.[53] The Navy Department concurred with these views.[54]

Returning now to the matter of the Bulgarian proposal, Bulgaria decided to withdraw its suggestion in favor of an Indian proposal, according to which the continental shelf was not to be used for the purpose of military bases or installations.[55] The expressed position of the United States was that the question of military bases went far beyond the subject matter considered by the International Law Commission; the United States opposed this amendment and it was defeated.[56]

As we have seen, the different departmental representatives included in the American delegation to the Geneva Conference brought with them their departmental preferences shaped by departmental perspectives. The Navy was vitally concerned with maintaining a narrow limit for territorial waters and preventing developments that might interfere with free transit in ocean areas. The State Department was concerned, to a great extent, with international fishery problems and the possibility of consequent interstate conflict. The Interior Department was motivated by the desire to preserve and extend coastal state control over offshore land. At this point in time, there was no perception that the various departmental preferences were in any significant way mutually exclusive.

NOTES

1. Arthur Dean, "Geneva Conference on the Law of the Sea," American Journal of International Law 52 (1958): 610.

2. For a list of members of the United States delegation at Geneva see A/Conf. 13/38, volume II, pp. XXV-XXVI.

3. This information is contained in a letter from Captain Hardy to the author.

4. On this point see Lt. Commander Keith D. Lawrence, "Military Legal Considerations in the Extension of Territorial Seas," Military Law Review 29 (1965): 92.

5. For a detailed list giving the least width of important straits and channels, see United States Department of State, Geographic Bulletin no. 3, Sovereignty of the Sea, Revised October 1969, pp. 22-28. Some states deny that warships have a right of innocent passage. This seems, for example, to be the view of the Soviet Union. See

William E. Butler, The Soviet Union and the Law of the Sea (Baltimore: The Johns Hopkins Press, 1971), p. 59.

6. Hardy letter, op. cit. and Dean, op. cit., p. 612.

7. Captain Robert D. Powers, Jr. and Captain Leonard Hardy, "How Wide the Territorial Sea?" in United States Naval War College, International Law Studies, 1959-1960 (Washington, D.C.: Government Printing Office, 1961), pp. 304-312.

8. Dean, op. cit., p. 608.

9. This information is derived from an interview in Washington, D.C. with a high ranking official long experienced in ocean affairs.

10. Hardy letter, op. cit.

11. 66 Stat. 883, approved August 27, 1954. This act was amended in 1968 to broaden the scope of reimbursement to include license fees, registration fees or any other direct charges actually paid to the foreign government. 82 Stat. 729, approved August 12, 1968. From 1954 to 1967, the total fines imposed against American ship owners by Latin American states amounted to $489,470. House Merchant Marine and Fisheries Committee, "Foreign Seizures of U.S. Fishing Vessels," 90th Congress, 1st session, p. 58.

12. This information is derived from an interview in Washington, D.C. This official pointed out that the political influence of the American fishery industry had substantially deteriorated by the middle 1960s.

13. A/Conf. 13/28, pp. 3-4.

14. This information is contained in a letter from an executive officer of the National Academy of Sciences to the author.

15. Department of the Interior, Bureau of Land Management, Information Service, for Release October 14, 1954 in the files of the Interior Department.

16. Department of the Interior, Bureau of Land Management, Information Service, for Release July 12, 1955 in the files of the Interior Department.

17. Information Service Release of October 14, 1954, op. cit.

18. A/Conf. 13/L.12. This wording adopted by committee IV was accepted in the plenary sessions and became the wording of article 1 of the Continental Shelf Convention.

19. A/Conf. 13/42, p. 19.

20. Ibid., p. 40.

21. Bernard Oxman, The Preparation of Article 1 of the Convention on the Continental Shelf (Washington, D.C.: U.S. Department of Commerce, National Bureau of Standards, 1968), p. 151.

22. Proceedings of the Fourth Annual Conference of the Law of the Sea Institute (Kingston: University of Rhode Island, 1970), p. 281.

23. Writing just after the conclusion of the Geneva Conference, Nicholas J. Campbell, Jr., Associate General Counsel of Standard Oil of New Jersey, expressed the view that the exploitability criterion is justified by the need to allow for future technological developments. See his article, "International Law Developments Concerning National Claims to and in Offshore Areas," Tulane Law Review 33 (1958/1959): 350. See also "Sea Mounds May Lead to Deep Water Oil Resources," Oil and Gas Journal, November 26, 1956, p. 58.

24. Oxman, op. cit., pp. 154-155.

25. Marjorie Whiteman, "Conference on the Law of the Sea: Convention on the Continental Shelf," American Journal of International Law 52 (1958): 636.

26. Ibid., p. 637; A/Conf. 13/SR. 8, pp. 8-9, 11.

27. A/Conf. 13/42, p. 17. This narrow view of "natural resources" of the shelf was shared by the representatives of Japan, Greece, and Belgium. See A/Conf. 13/42, pp. 56, 58-59, 63.

28. A/Conf. 13/42, p. 28.

29. See, for example, the statements of the representatives of India, A/Conf. 13/42, p. 12; Mexico, A/Conf. 13/42, p. 15; Pakistan, A/Conf. 13/42, p. 19; United Arab Republic, A/Conf 13/42, pp. 27-28; Ireland, A/Conf. 13/42, p. 29; and Yugoslavia, A/Conf. 13/42, p. 50.

30. 559 UNTS 285.

31. A/Conf. 13/42, p. 12.

32. Chapman, "Conflict of Uses of the Sea," Proceedings of the Second Annual Conference of the Law of the Sea Institute, op. cit., pp. 70-71.

33. Arthur Dean has written that the doctrine of "abstention" is "to the effect that where a nation or nations have invested time and capital in the development of a fishery in a particular stock of fish, or area of the sea, nations which have not formerly fished that stock, or have not contributed to the development of the area, should abstain from fishing there in the future. The doctrine is one of conservation and of equity." Dean, "Geneva Conference on the Law of the Sea," op. cit., pp. 626-627. For a detailed analysis of the principle of abstention see Richard Van Cleve, "The Economic and Scientific Basis of the Principle of Abstention," A/Conf. 13/3.

34. A/Conf. 13/42, p. 20.

35. Letter from Stillman Wright of the Fish and Wildlife Service, op. cit., p. 6. This fact was also pointed out in a preparatory study for the conference prepared by the Food and Agricultural Organization. The study indicated that "in nature there is no simple line of demarcation between sedentary and other fish, but only a long series of gradations from the unquestionably fixed at one extreme to the unquestionably free at the other." A/Conf. 13/13.

36. A/Conf. 13/42, p. 62.

37. Ibid.

38. Testimony of Arthur Dean before the Senate Foreign Relations Committee, "Conventions on the Law of the Sea," 86th Congress, 2d session, p. 88.

39. John Knauss in *Proceedings of the Fourth Annual Conference of the Law of the Sea Institute*, op. cit., p. 400.

40. A/Conf. 13/42, p. 88.

41. A/Conf. 13/C. 4/L. 49.

42. A/Conf. 13/C. 4/SR. 28, pp. 4-5; A/Conf. 13/C. 4/SR. 30, p. 9.

43. *Proceedings of the Fourth Annual Conference of the Law of the Sea Institute*, op. cit., p. 403.

44. Ibid., p. 397.

45. Letter of D. W. Bronk to the Secretary of State, dated January 28, 1959. This letter was found in the files of the Fish and Wildlife Service, Federal Records Center, Suitland, Maryland. Access number 67 A 1358 box 80.

46. Letter of Ross Leffler, Assistant Secretary of the Interior, to Loftus Becker, Legal Adviser, Department of State, dated April 1, 1959, Ibid.

47. Milner Schaefer, "The Changing Law of the Sea—Effects on Freedom of Scientific Investigation," in *Proceedings of the Second Annual Conference of the Law of the Sea Institute*, op. cit., pp. 113-120.

48. A/Conf. 13/C. 4/L. 41/Rev. 1.

49. A/Conf. 13/42, pp. 83-84.

50. See the New York *Times* of April 10, 1955, p. 28; May 21, 1955, p. 7; and September 29, 1955, p. 7.

51. Hanson Baldwin, "Our Radar Ramparts," *New York Times Magazine*, November 13, 1955, p. 11.

52. New York *Times*, August 2, 1956, p. 49. Two other such radar stations were placed about 75 miles east of New York and about 90 miles east of Portland, Maine. New York *Times*, September 29, 1955, p. 7.

53. Letter from Colonel Walter D. Reed, Chief, International Law Division, Office of the Judge Advocate General, Department of the Air Force to the author.

54. Letter from Captain John R. Brock, Judge Advocate General's Office, Department of the Navy, to the author.

55. A/Conf. 13/C.4/L.57.

56. A/Conf. 13/42, pp. 38, 91.

CHAPTER

5

THE EARLY 1960s

INTERIOR DEPARTMENT ATTITUDES AND POLICY

The United States Department of the Interior, the operating executive agency with day to day responsibility for American policy on the continental shelf, seemed to be well satisfied with the 1958 Continental Shelf Convention. In 1960 it was pointed out by Theodore Stevens, the solicitor of that department, that the Senate had just given its consent to the Continental Shelf Convention, thereby recognizing that the coastal state had jurisdiction over the seabed "adjacent to its shores extending, as a minimum, to the point where the water depth reaches 200 meters, and beyond that to whatever distance minerals or other valuable resources therein are exploitable."[1] It is interesting to note that although the criterion of adjacency is mentioned, it appears to be effectively overridden by the observation that exploitation is allowed "to whatever distance minerals or other valuable resources are exploitable." The Interior Department thus stressed the elastic nature of the exploitability clause, which allowed for the maximization of the area that could be brought under the jurisdiction of the United States and, in particular, the jurisdiction of the Interior Department.

It was not long before the ideas expressed by Interior Solicitor Stevens were practically applied. In 1960 the Collier Carbon and Chemical Corporation filed a request with the Bureau of Land Management for an application to lease a parcel of land off the coast of California for phosphate exploitation. This land was located at a distance of 40 miles off the California coast in a water depth that ranged between 240 and 4,000 feet. In addition, between this area and the mainland there was a deep channel greater than 3,600 feet in depth.[2] Although the Interior Department encouraged the filing of this request,[3] the director of the Bureau of Land Management was not sure that he had the authority to issue a lease in that area. In a letter to the associate solicitor he observed that the continental shelf is generally defined as extending into the sea to the point where the water rapidly deepens; this point usually occurs at a water depth of about 600 feet. The land that the Collier Company wanted to lease was well beyond this depth and was said to be "a part of a plateau rising from the ocean floor."[4]

In an important opinion, the Interior Solicitor's Office replied to this memorandum affirming the right of the Bureau of Land Management to grant such a lease. That the area under consideration was 40 miles from the mainland was said to be of little importance, since in the Gulf of Mexico, where the depth of the water increases very slowly, activities sanctioned by the Outer Continental Shelf Lands Act had been conducted at even greater distances from the shore. As to the outer limit of the continental shelf, the opinion observed that "at the time when the Outer Continental Shelf Lands Act was passed the Congress did not contemplate the immediate development of areas beyond the 100-fathom line." It was asserted, however, that Congress, in passing the act, did not prescribe a limit to the extent of the American continental shelf. It was claimed that "without such a seaward limit it is possible to hold that with technological progress this act may be applied to areas at greater and greater depths."[5]

Whether this position could be strongly maintained in the light of the legislative history of the Outer Continental Shelf Lands Act is at best questionable; this assertion, however, is interesting insofar as it indicates the preferences of the Interior Department relative to offshore areas, The Solicitor's Office, though, did not base its opinion to the Bureau of Land Management on the Outer Continental Shelf Lands Act alone. That office further observed that on May 26, 1960, the Senate "ratified" the Continental Shelf Convention.[6] It was stated that this convention, though not yet in effect, could be regarded as expressing the present views of the United States on the continental shelf.

Article one of the convention containing the definition of the continental shelf was then quoted. According to the Solicitor's Office, this was the first definition of the "continental shelf" accepted by the United States government that set a seaward limit. The Outer Continental Shelf Lands Act was applicable to all of the submerged land that appertained to the United States; since this country had asserted rights to the seabed to the limit of exploitability, the act then applied to all the land within this limit. Consequently, the Solicitor's Office ruled that the land which the Collier Company wished to lease could, indeed, be leased in accordance with this act.[7] Before being issued this opinion was submitted to the State Department and to the Justice Department for evaluation; no objections were expressed by either department.[8]

The opinion of the Solicitor's Office ignored one important point and dealt with another significant point only by implication. Nowhere in the opinion was the criterion of adjacency, also found in article one of the Continental Shelf Convention, considered. The accent of the opinion was exclusively on the matter of exploitability. Although the opinion indicated that the convention's definition was the first that set any seaward limit to the shelf, the tenor of the opinion suggested a limit constantly moving further off the coast as technology develops and exploitability becomes possible in deeper waters. Additionally, the fact that a deep submarine trench of at least 3,600 feet separated the area under consideration from the mainland was not explicitly considered. The implication of this opinion was that a state's jurisdiction could bound over and across a submarine chasm and extend to the land on the seaward side that was exploitable. This legal opinion and subsequent leasing in the area could be cited as precedents and used as guides by other states, such as Norway, whose nearby offshore areas contained deep submarine trenches.

To the extent that Congress at this time gave it thought, it seemed to support and encourage the expansive tendencies of the Interior Department. In July of 1960 the House Science and Astronautics Committee concluded that while the oceans and their contents in the areas beyond territorial waters were owned in common (res communis), the ocean bottom, as a consequence of the Continental Shelf Convention, was "everywhere defined as belonging to the coastal State with the median boundary instantly defined, everywhere, sometimes existing at mid-ocean."[9] In stating this belief, the committee went beyond any claim made by the Interior Department. Accordingly, this view could well serve to encourage the expansionist tendencies already existent in the Interior Department. Insofar as could be learned, no comment was made by any official of the executive branch on the stated view of the Science and Astronautics Committee.

Further encouraging the leasing of more distant offshore land was the desire of the Interior Department to make a favorable impression on the Appropriations Committees of Congress.[10] It is a fact that the leasing of offshore lands produces the greatest receipts of any Interior Department activity. A sustained or, better yet, increased program of leasing with resultant increased bonuses and royalty payments would reflect well upon the efficiency and importance of the Department of the Interior. A decrease in that department's income from offshore leasing might possibly raise questions as to the functioning of the Interior Department. This is seen, for example, in the questioning of Karl Landstrom, Director of the Bureau of Land Management, by Congressman Denton during hearings held by the House Appropriations Committee in 1963. Congressman Denton wanted to know why the estimated total receipts of the Interior

Department for 1964 would be down from the figures for 1963. Landstrom blamed the decrease in income on the offshore situation; he pointed out that while the royalties of existing wells were increasing, the bonuses that Interior was receiving as it leased in deeper waters were declining to some degree.[11] To forestall criticism of the type implied by Congressman Denton, it could be expected that the Interior Department would attempt to lease even more offshore land in an attempt to make up for the smaller bonus payments.

In fact, a policy of this sort did not suit all of the petroleum companies. At this time the independent oil producers were becoming very concerned about the steadily increasing outputs from offshore areas and their effect upon petroleum prices. Additionally, there was strong feeling among the independent producers that the offshore sales benefited only the giant petroleum companies. Though leases were granted through a system of competitive bidding, only the largest companies could afford to obtain them. Often so much capital was needed that even the large companies found it necessary to carry out joint operations.[12] Despite opposition from independent producers in 1962, for example, the Interior Department went ahead with its leasing plans in the Gulf of Mexico where it expected to realize $500,000,000 that year.[13] Independents opposed to this sale pointed to the tremendous excess capacity said to be already burdening the oil industry and suggested that the oil in the Gulf of Mexico be kept in place as a reserve.

In an attempt to persuade the Interior Department to postpone the lease sale scheduled for April of 1962, Congressman Jim Wright of Texas contacted that department in March on behalf of the independent producers. He stated his belief that "every effort should be made to get the domestic on-shore oil business in a stronger, healthier condition before encouraging the development of the outer continental shelf."[14] Congressman Wright's request for a postponement was turned down by Assistant Interior Secretary John Kelly, who claimed that the interests of the United States were best served by proceeding with the sale as scheduled. Kelly expressed the view of the Interior Department that "it is necessary to develop the technology required to explore, drill, develop, and produce oil from extreme water depths in the outer Continental Shelf."[15]

For their part, the large oil companies eagerly sought offshore leases and called for increases in offshore prospecting. In 1966, for example, Frank Ikard, President of the American Petroleum Institute, the trade association of the largest oil companies, called for a continued increase in offshore petroleum exploitation.[16] Testifying before a congressional committee in 1965, a spokesman for the API observed that future years held every prospect for expansion of oil and gas exploitation in an ever increasing offshore area. In this context he noted that American jurisdiction extended to the limit of exploitability even if beyond the continental shelf itself. In such an area, the API claimed, "existing international law, which provides for us to move outward from the Continental shelf should pertain."[17] Thus, the tendencies that existed within the Interior Department to lease in more distant offshore areas were reinforced by the desires of an important Interior clientele group, the larger petroleum companies;[18] a convergence of interests existed.

NAVY CONCERN WITH INTERIOR DEPARTMENT POLICY

In addition to the legal problems that were engendered by the request of the Collier Corporation, questions were raised in regard to such leasing by the United States Navy. The Navy made it known that it would not object to the granting of a lease in the area desired by the Collier Corporation if the lease were issued with the stipulation that the lessee would agree to remove all his equipment, at his own expense, if, in the opinion of the Navy, the defense interests of the United States required such removal.[19] Recognition of conflict between different uses of ocean space had occurred; Navy officials were of the belief that further offshore leasing off California would interfere with naval operations and defense plans.[20] The Department of the Interior suggested that the legitimate interests of the Defense Department could be protected by a lease stipulation patterned after section 12 (c) of the Outer Continental Shelf Lands Act. This portion of the act reads:

> All leases issued under this Act, and leases, the maintenance and operation of which are authorized under this Act, shall contain or be construed to contain a provision whereby authority is vested in the Secretary [of the Interior], upon a recommendation of the Secretary of Defense, during a state of war or national emergency declared by the Congress or the President of the United States after the effective date of this Act, to suspend operations under any lease; and all such leases shall contain or be construed to contain provisions for the payment of just compensation to the lessee whose operations are thus suspended.[21]

This section protects the financial investment of offshore operations while the Defense Department stipulation could involve substantial costs with the concommitant effect of reducing the economic desirability of offshore exploitation. The wording of section 12 (c) also requires that certain conditions, beyond the control of the Navy, that is, a declaration of war or national emergency, be met before offshore operations are suspended. Additionally, nothing is said of removal of offshore equipment. Further, the wording of this section makes the Defense Department dependent upon the decision of the Interior Secretary even when conditions that would provide sufficient cause for suspension of commercial activities were present.

It is thus not surprising that Navy officials declined to follow the suggestion of the Interior Department.[22] Growing Interior Department impatience with the delay caused by the Navy with respect to the Collier phosphate lease is reflected in the letter of Interior Secretary Udall to Defense Secretary McNamara dated June 1, 1961. The very fact that this matter had to be dealt with at the secretarial level indicates its importance. In this letter Udall pointed to the Interior Department's responsibility to lease those offshore areas in which sufficient interest had been manifested. It was said that the conditions that the Navy wished to impose on Interior would seriously impede that department's efforts and could totally prevent the development of valuable mineral resources in the outer continental shelf. The Interior Secretary suggested that the Navy reconsider its position and accept a lease provision consistent with section 12 (c) of the Outer Continental Shelf Lands Act.[23]

Despite the intervention of Secretary Udall himself and despite an interagency conference held in August of 1961, the differences between the Navy and the Interior Department were not reconciled. The Navy held to its position on the lease stipulation and made it clear that it wanted all Pacific coast leases reviewed by it on a case by case basis prior to Interior leasing.[24]

In the face of the continuing intransigence of the Defense Department, the Department of the Interior acted to break the interagency deadlock. In November of 1961, it notified the Department of Defense that it was going ahead with the Collier phosphate lease subject to one of the conditions suggested by the Navy, namely a provision for the suspension of operations if in the interest of national security. It was noted that this condition was accepted only to avoid any further delay in granting the Collier lease and it was firmly stressed by Interior that this action "should not be construed as establishing a precedent in future similar lease offerings."[25]

On February 1, 1962 the Interior Department granted the Collier Company its long awaited lease. The text of the lease contains the standard reference to section 12 (c) of the OCS Lands Act; the concession to the Navy is found in a lease stipulation which reads:

> The right to suspend temporarily operations of the lessee under this lease in the interests of national security. Such temporary suspension of operations will come into effect upon the order of the cognizant U.S. Navy Sea Frontier Commander, when determined by the said Commander, or higher authority that national security interests necessitate such action. During any ordered period of temporary suspension of operations all payments of rentals, minimum royalty, and royalty prescribed by said lease shall likewise be suspended during such period of suspension and there shall be added to the term of said lease a period equal to the period of suspension.[26]

Removal of the lessee's equipment is not provided for in the lease and the lessee is protected from undue hardship by the terms of this stipulation. Clearly, the Interior Department was attempting to maintain to the maximum degree the economic attractiveness of offshore leasing.

The response of the Defense Department to Interior's action was neither favorable nor passive. Assistant Defense Secretary T. D. Morris observed that the responsibility of his department "for national defense and security measures involving highly sensitive military operations conducted in the Outer Continental Shelf areas cannot be carried out under conflicting conditions which would prevail if unrestricted leasing practices are permitted." Morris asserted that it was essential for certain portions of the shelf to be closed to all commercial operations and that leases granted in all other areas contain stipulations necessary to prevent interference with Defense Department operations.[27]

The Interior Department strongly opposed these demands noting that to accede to them "would substantially reduce the acreage which the oil and gas industry has indicated an interest to lease, and would undoubtedly result in an appreciable loss to the United States in bonuses and lease rentals, and in royalties should the area contain valuable oil and gas deposits."[28] Consequently, the Interior Department concluded that it would not be in the best interest of the United States to include in offshore leases conditions not provided for in the OCS Lands Act. Such lack of cooperation from the Interior Department disturbed the Defense Department and contributed to the development of Defense attitudes on the continental shelf; this experience could not help but create ill feeling and suspicion between these two executive departments.

NAVY INTERESTS

The growing concern of the United States Navy in oceanography and the submarine floor during the late 1950s and early 1960s was the direct result of increasing preoccupation with the threat posed to American security by the Soviet submarine fleet. In order to counter this threat and to develop the technology and methods of effective antisubmarine warfare (ASW), increased attention had to be paid to the physical characteristics of the ocean environment. In 1960, the House Committee on Science and Astronautics issued a report entitled "Ocean Sciences and National Security" in which it expressed deep concern with Soviet submarine capability. While the committee found that there were a variety of reasons why oceanic research should be conducted, it observed that the most compelling stimulus for such research was purely military. The committee concluded its report by indicating the desirability of developing the means for complete surveillance of the oceans so as to minimize the possibility of surprise attack from beneath the sea.[29]

The concern of the Science and Astronautics Committee was shared by Navy officials. In 1959, noting the threat to American security emanating from the growing Soviet submarine fleet, Admiral Arleigh Burke, Chief of Naval Operations, urged greater research in the field of antisubmarine warfare. The primary problems that had to be overcome to defeat the submarine were said to be those of detection, identification, and accurate localization. According to Admiral Burke significant advances in detection systems were beginning to emerge from Navy research and development.[30]

The Chief of Naval Operations may well have been referring to Project Caesar research, which had been going on through the 1950s. The Caesar ASW system became fully operational in 1964 after the expenditure of approximately $160,000,000. It involved the placing of a complex system of linked hydrophones on the United States continental shelf in the Atlantic Ocean at a depth of 600 feet powered by onshore generators and linked to a shore based computer system which was able to interpret rapidly collected data.[31] As of 1965, it was reported that the Caesar system could detect and locate submarines approaching to within a few hundred miles of the coast of the United States.[32] That the Navy was pleased with the operation of this system was seen in its plan to install a similar system off the Pacific coast of the United States.[33]

As Navy interest increased in ASW and the development and placing of sophisticated submarine tracking gear, the potential for conflict with the Interior Department began to grow. Effective coastal tracking systems required the freedom to deploy acoustic devices as need demanded, yet commercial continental shelf installations could well interfere with the placing and operation of these tracking devices. The practice of the Interior Department in granting leases to areas farther and farther offshore could be expected to be followed by other states. This development could make the ASW task of the United States Navy more difficult.

Dr. John Craven, formerly of the Navy Special Projects Office, has said that the easiest way to track a submarine is to "pick it up" with listening devices as it leaves its home port and to follow its movements continuously. He notes that once an undetected submarine gets to the open sea the task of then detecting its location is very much more complex and difficult.[34] Because of the relative ease with which a submarine could be tracked as it leaves port, antisubmarine forces could naturally be expected to consider the planting of acoustic tracking devices on the continental margin of other states. To counter the possibility that its submarines could be trailed from the inception of their voyages, Dr. Craven suggested that a state such as the Soviet Union would be strongly tempted to push the development of international law in such a way that a safe perimeter would be established around its coasts.[35] In such a zone other states might be prevented from planting tracking devices. The zone would be of such an extent that submarines could lose themselves in oceanic waters without any trail being established. At this point antisubmarine forces would be in the proverbial position of finding a needle in a haystack.

Since antisubmarine efforts were of great importance to the Navy, it would seem that the development of exclusive national control over wide areas of the sea bottom adjacent to coastal states would be opposed to Navy interests. A tendency toward wider areas of seabed under national control was in fact being encouraged by the leasing policies of the United States Interior Department. Why during this period did the Navy not take a firm position and insist that the Interior Department provide some limit beyond which it would not lease offshore? With regard to surface mobility, the concept of "creeping jurisdiction" was not yet fully developed in Navy thought. Navy thinking at the time of the Geneva Conference on the Law of the Sea, was that offshore land could be dealt with apart from the matter of the status of superjacent waters. But what about the growing desire to develop a more effective ASW effort through the placing of tracking devices close to the shores of foreign states? To a great extent this consideration was balanced by the possibility of new weapons systems whose requirements were perceived to be opposed to those of potential ASW systems.[36]

In terms of non-ASW sea bottom systems, a panel of the President's Science Advisory Committee suggested in its 1966 report the possibility of emplacement of missiles on underwater barges on the floor of the continental shelf. Such barges could from time to time be moved to conceal their location. Also considered was a mobile ocean bottom system, which could creep along the floor of the sea.[37] That the Navy was interested in learning more about this type of weapon system is reflected in their granting of a contract in 1964 to Westinghouse for studies of a bottom mobile weapon system.[38]

Another potential military use of the seabed considered at this period involved the planting of fixed missile silos directly into the continental shelves.[39] As late as October 1967, Robert Frosch, Assistant Secretary of the Navy for Research and Development, stated that "Future design of sea based deterrents following Polaris/Poseidon may take many forms. Underwater silos, for example, are a possibility."[40]

A third possible use of the seabed involved the construction of American submarine bases in the continental shelves. Writing in 1968, Professor William Nierenberg noted that there was indeed a possibility that submarine bases would be built in shelf areas and linked to shore with a variety of communications and transport facilities.[41] This idea was not so far beyond the realm of technical possibility, since as early as 1963 an American architectural engineer patented an undersea island he had designed. The structure was built to house a naval base or an offshore drilling rig; for use as a submarine base the design provided for the berthing of submarines and access hatches. The United States Navy was known to be interested in this invention.[42] In fact, the Navy was studying proposals of this type on its own under the code name of Project Rocksite. This undertaking was concerned with the drilling of tunnels into the seabed and the combining of existing capability of forming mine tunnels and shafts beneath the sea floor with a new technology for mating submersibles to a sea floor shaft entrance, thus providing a completely independent subsea floor installation.[43]

Military systems of the type described above carried with them the necessity for exclusive control by the coastal state over wide sea bottom areas so that secrecy and security could be maintained. In the last analysis, however, the Defense Department decided against the development of these weapons systems. The costs of such systems were felt to be prohibitive; one scientist has claimed that their cost would make the space program budget look miniscule by comparison.[44] The decision against the development of these systems opened the way for a clear Defense Department position favoring a narrow continental shelf under coastal state jurisdiction.

FISHERY INTERESTS

The history of the American fishing industry in recent years has been marked by stagnation and decline. While the fishing fleets of other states such as those of the Soviet Union and Japan were new and used the most modern electronic and mechanical gear, American fishermen continued to make use of old and obsolete equipment.[45] The decline of the American fishing industry is evidenced by statistics. In 1948 American fishermen supplied nearly 13 percent of the world's fishery production; by 1962 that figure had decreased to 7.4 percent. In the decade from 1950 to 1960, the number of Americans working as commercial fishermen dropped from 161,000 to 130,000; the number of fishing vessels decreased by 16,000 during this same period. In 1950 American fishermen caught almost 80 percent of the fish consumed in this country; by 1961 this figure had been reduced to 56 percent.[46]

A major factor contributing to the decline of American fishing was the appearance of foreign fishing fleets off the coasts of the United States. Until 1958 fishing grounds close to American shores were rarely subject to operations by foreign fishermen, but soon after that the situation began to change very radically. The Soviet Union, for example, had undertaken a rapid development of its fishing fleet and it increasingly sought to exploit fishery resources on a large scale. From 1958 to 1962 the catch of Soviet fishermen jumped from 2,621,000 metric tons to 3,616,000 metric tons. During the same period the catch of American fishermen crept from 2,708,000 to 2,904,000 metric tons.[47] More significant than the mere fact of rapid growth on the part of the Soviet fishing industry is the fact that Soviet fishing vessels began to operate in very substantial and ever increasing numbers off coastal areas of the United States. Soviet ships began to fish off the United States coast in the Bering Sea in 1959, off New England in 1961, in the Gulf of Alaska in 1962, in the Gulf of Mexico in 1962-1963, off Oregon and Washington in 1966, off California in 1967, and off Hawaii in 1968.[48]

A strong reaction from American fishermen to the developing threat from the Soviet fishing fleets was not long in coming. They contacted their representatives and senators in Congress and urged that something be done about the Soviet fishing fleets operating beyond the three-mile limit of territorial waters but in areas considered by American fishermen to be traditional American fishing grounds. Members of Congress, in turn, called upon the executive branch to take action. In September of 1963, dissatisfied with what it felt to be inaction on the part of the executive branch, the Senate Commerce Committee, Subcommittee on Merchant Marine and Fisheries, began hearings on "Fishing in U.S. Territorial Waters." The title of these hearings is misleading since ocean areas beyond the limit of territorial waters were actively examined.

The committee considered a bill introduced by Senator Bartlett of Alaska, which would make it unlawful for foreign ships to take fish from American territorial waters or to take continental shelf fishery resources that belonged to the United States.[49] William C. Herrington, who testified on the Bartlett bill on behalf of the State Department, noted that department's support for the purposes of the bill but wondered if reference to the continental shelf was not "premature." His concern centered on the problem of enforcement of the continental shelf provision against foreign ships on the high seas; it was suggested that this was a matter that required further study since international law did not seem to provide a basis for such enforcement.[50] For now he felt that mention of the continental shelf should be omitted.

If this provision were not to be dropped, Herrington urged that it be made clear that it should apply only to living resources that appertain to the United States pursuant to the Geneva Convention on the Continental Shelf.[51] The State Department wanted it understood that should the Bartlett bill become law, foreign fishing in waters over United States continental shelf areas was to be limited only in accordance with the internationally agreed upon Continental Shelf Convention; the bill was not to apply to foreign fishing operations over American continental shelf areas generally. It is evident from the testimony of Herrington that the State Department was deeply concerned by possible damage that might befall the American distant water tuna and shrimp industries. If fishing rights for foreign states over United States continental shelf areas were too greatly restricted these important American distant water fisheries could suffer through action taken by other states.[52]

While the Interior Department agreed that it might be wise to delete the continental shelf provision, it strongly recommended that a section be added to the bill which would give the Interior Department the power to define exactly what species were to be considered as living resources of the continental shelf. Interior Secretary Udall noted that continental shelf fishery resources had nowhere been authoritatively defined and that no procedure had been yet established whereby foreign fishermen could be advised as to which fishery resources were denied to them.[53]

The Bartlett bill did become law on May 20, 1964 and in its final form contained the addition suggested by the Interior Department.[54] It was not, however, until late October of 1968 that the Secretary of the Interior, after consultation with the Secretary of State, made a determination as to what species of living organisms constituted continental shelf fishery resources and thus were to be exploited exclusively by American nationals.[55]

Elements within the American fishing industry wanted Congress to take action of a stronger nature than that suggested in the Bartlett bill. Congressman Keith noted that he had been urged by the Seafood Producer's Association of New Bedford to support legislation that would keep foreign fishing vessels away from American shores to the extent of 200 miles.[56] James Ackert, of the Atlantic Fishermen's Union, observed that from January 1958 to July 1958 the Boston fishing fleet produced 81.7 million pounds of fish; for the same period in 1963 the figure was 66.7 million pounds. He placed the blame for this significant decrease in catch upon the Soviet fishing fleet, which operated off New England. His simplistic solution to the problem was to ban foreign fishing vessels from operating in the waters above the continental shelf of the United States.[57]

The desire to extend the exclusive fishery jurisdiction zone of the United States in such a manner was a nightmare for American shrimp interests and underlined the fact that the needs and requirements of the various United States fisheries were different. According to William Neblett, Executive Director of the National Shrimp Congress, approximately 20 percent of the catch of American shrimp fishermen came from waters off the coasts of other states. American shrimpers feared that United States exclusion of foreign fishermen from waters over American shelf areas would be reciprocated. Since shrimp were found in water depths ranging from 10 to 50 fathoms, areas which would be considered as shelf areas, the American shrimp industry, with its significant distant water operations, stood to suffer very substantial losses. The additional point was made that any reduction of American shrimp fishing off foreign coasts would inevitably result in an "impossible concentration" of too many American vessels on home grounds leading, in turn, to "complete economic disaster" for the American shrimp industry. While sympathetic to the plight of American coastal fishermen, Neblett cautioned the Commerce Committee that in devising measures to protect one fishery it should not coincidentally destroy another. He very strongly urged that legislation should not connect the continental shelf to fisheries unless it very clearly reaffirmed the language of the Continental Shelf Convention, which limited the living resources under the coastal state's exclusive jurisdiction.[58]

In 1966 the United States Congress did take the more modest step of extending exclusive American fishery rights to 12 miles from the coast.[59] While in 1962 the State Department had felt that international law provided no basis for a zone of such extent,[60] by 1966 it suggested that in view of recent developments in international practice such a claim would no longer be contrary to international law.[61]

During 1966 attempts had been made in Congress to extend exclusive fishing rights to the seaward boundary formed by the 200-meter depth contour. Bills to this effect were introduced by Congressmen Pelly and Watt; in the words of Congressman Downing, who chaired the hearings on these bills, they "would have the effect of establishing an exclusive fisheries zone superjacent to the Continental shelf."[62] Representatives of executive agencies testifying on these bills were unanimous in strongly opposing them. Spokesmen for the Commerce and Interior Departments claimed that on economic and commercial grounds no advantage could be gained from the enactment of this type of legislation. Noting the importance of the American tuna industry, which fishes off the western coast of Latin America, it was suggested that these measures could have significant adverse impact on that industry.[63] Representatives of both the State Department and the Defense Department observed that a claim such as that embodied in the proposed legislation would violate international law.[64] Additionally, the Defense Department showed concern over possible indirect effects on navigation rights on the high seas. Emerging anxiety with "creeping jurisdiction" was reflected in the statement of Commander Downs on behalf of Defense that, "We are also concerned that illegal action to claim exclusive fishing rights in large areas of the high seas may accelerate a trend of claims which will seriously interfere with freedom of navigation on and over the high seas."[65]

THE KING CRAB PROBLEM

Pressure to have king crab declared a continental shelf resource dates back to at least 1963. In that year Senator Bartlett urged that king crab be protected under the terms of the Continental Shelf Convention.[66] Historically, the Japanese had been taking king crab in the Bering Sea area before American king crab fishery began. By 1959 the Soviet Union was also capturing king crab in this area. Of the 7,980,000 king crabs caught in the Eastern Bering Sea in 1962 only 10,000 were captured by American fishermen; the remaining crabs were taken by Japanese and Soviet fishermen.[67] While American fishermen made use of traps to capture king crab, the Japanese and Soviets employed tangle nets, equipment which was forbidden to United States fishermen. Eventually, due to the intensive operations of the foreign fishermen, the king crab population began to decline.[68]

Despite the concern of American fishing interests, it was not until 1968 that king crab was classified as a living resource of the continental shelf. The explanation for this delay is instructive. Basically, two points are involved, one concerning the need for scientific study and the other the need for diplomatic negotiations.[69] First, careful technical studies had to be done before a species could be judged immobile on or under the seabed or unable to move except in constant physical contact with the seabed or subsoil. It was of very great importance to provide a firm scientific basis for declaring a living species a continental shelf resource. It was feared that if the United States did not establish a clear and justifiable scientific rationale for an American claim, other states could make claims of their own, detrimental to American interests, on similarly loose grounds.

Second, a mere unilateral declaration by the United States claiming king crab as a continental shelf resource might not have been recognized by other states. Since, in the case of king crab, a declaration of this type would directly affect nationals of Japan and the Soviet Union there was a strong desire within the United States government to reach an understanding with these states. According to David Windley of the Bureau of Commercial Fisheries, bilateral negotiations with the Japanese and Soviet governments began in late 1964 and early 1965 respectively. With regard to the talks between the United States and the Soviet Union, both states agreed that king crab was a continental shelf resource. It was also agreed that a sudden end to the Soviet king crab fishery would cause economic hardship to the Soviet Union and that king crabs were not fully exploited by American fishermen.[70] In this context a Soviet-American agreement on king crab was signed on February 5, 1965.[71] In accordance with this agreement, the United States allowed the Soviet Union to continue taking king crab in United States continental shelf areas subject to: 1) reduced catch, 2) reciprocal rights of boarding king crab fishing vessels to observe enforcement of the agreement, 3) exchange of scientific data on the exploited stocks, 4) delimitation of an area where king crabs could be fished only by crab pots, and 5) no Soviet king crab fishing south of the Aleutian Islands. This agreement with the Soviet Union has been renegotiated several times since 1965 with the Soviet Union each time accepting further reductions in its allowable catch and increased areas in which only crab pots could be used.[72]

With regard to Japan greater obstacles existed. Unlike the Soviet Union, Japan was not a party to the Continental Shelf Convention; moreover, Japan maintained that king crab was a high seas fishery resource and thus not subject to the exclusive use and control of the United States.[73] Despite their differences, the two states did sign a king crab agreement on November 25, 1964.[74] Under the terms of this agreement Japan agreed to reduce its catch of king crab and to provide scientific data on the species. Furthermore, an area was delineated in which only crab pots could be employed. This agreement was also renegotiated several times; in each case Japan agreed to a further reduction in catch and to an increased area in which only crab pots were to be used.[75]

The United States government had learned from experience that there were distinct disadvantages to unilateral action with respect to claims made to ocean areas. It had delayed its assertion to exclusive rights to king crab in order to assure the effectiveness of its claim, to minimize friction with other important states, and to underline, through careful technical studies, the scientific basis for the American claim to king crab. As projected, the American king crab claim was to be shown as having a rational and legitimate basis and as being internationally agreed upon as opposed to being an arbitrary unilateral act of resource expropriation. There was to be no encouragement given to the general and extensive unilateral fishery claims of Latin American states.

NOTES

1. Letter of Interior Solicitor Stevens to Mr. Bill Shiels, dated November 13, 1960. In the files of the Interior Department, Minerals and Fuels branch, Federal Records Center, Suitland, Maryland.

2. Letter on behalf of the Collier Carbon and Chemical Corporation from the law firm of Brobeck, Phleger and Harrison to the Director, Bureau of Land Management, dated August 18, 1960. Ibid.

3. A letter signed by A. J. Rockwell of Brobeck, Phleger and Harrison and addressed to C. R. Bradshaw, Associate Solicitor for Public Lands, Interior Department, notes that the Collier request "was prepared in the light of the conference held in your office. . ." The letter was dated August 19, 1960. Ibid.

4. Letter from the Director of the Bureau of Land Management to the Associate Solicitor, Division of Public Lands, dated April 13, 1961. Ibid.

5. Thomas J. Cavanaugh, Associate Solicitor, Division of Public Lands, "Application of Outer Continental Shelf Lands Act to Designated Area Off the Coast of California," M-36615, dated May 5, 1961.

6. The term "ratified" is here being used incorrectly. As Hackworth notes the Senate does not ratify treaties but instead consents to their ratification by the President. Digest of International Law 5 (1943): 48.

7. Cavanaugh opinion, op. cit.

8. See the remarks of Frank J. Barry, Solicitor, Department of the Interior, "Administration of Laws for the Exploitation of Offshore Minerals in the United States and Abroad," at the American Bar Association, National Institute on Marine Resources, Long Beach, California on June 9, 1967.

9. House Report No. 2078, "Ocean Sciences and National Security," 86th Congress, 2d session.

10. On the importance of the relationship between executive agencies and congressional appropriations committees see Aaron

Wildavsky, The Politics of the Budgetary Process (Boston: Little, Brown and Company, 1964), pp. 47-62.

11. House Appropriations Committee, Subcommittee on the Department of the Interior and Related Agencies, "Department of the Interior and Related Agencies, Appropriations for 1964," 88th Congress, 2d session, p. 103.

12. New York Times, July 5, 1964, section 3, p. 1.

13. New York Times, February 25, 1962, section 3, pp. 1, 28, and February 19, 1962, p. 35.

14. Letter of Congressman Wright to Interior Secretary Udall, dated March 2, 1962. In the files of the Department of the Interior, Office of the Secretary—Submerged Lands.

15. Letter of Assistant Interior Secretary Kelly to Congressman Wright, dated March 9, 1962. Ibid.

16. New York Times, June 28, 1966, p. 16.

17. See the testimony of Durland E. Clark, Jr. before the House Merchant Marine and Fisheries Committee, Subcommittee on Oceanography, "National Oceanographic Program Legislation," 89th Congress, 1st session, p. 376.

18. On the important political position of the large oil companies see Robert Engler, The Politics of Oil: A Study of Private Power and Democratic Directions (Chicago: University of Chicago Press, 1967); and Erwin Knoll, "The Oil Lobby is Not Depleted," New York Times Magazine, March 8, 1970.

19. Letter of Max Caplan, Minerals Staff Officer, Department of the Interior, to Rear Admiral F. L. Johnson, dated March 27, 1961. In the files of the Interior Department, Minerals and Fuels branch, Federal Records Center, Suitland, Maryland. Access number 69 A 2058 box 90.

20. Memorandum for the record of Karl Lanstrom, Director of the Bureau of Land Management, regarding the conference held at the Pentagon on August 15, 1961, dated August 16, 1961. Ibid.

21. Public Law 212, 83rd Congress, 1st session.

22. Letter from Rear Admiral Johnson to Max Caplan, Department of the Interior, dated May 5, 1961. In the files of the Interior Department, Federal Records Center. Access number 69 A 2058 box 90. Ibid.

23. This letter is found in Ibid.

24. Letter of T. D. Morris, Assistant Secretary of Defense to Assistant Interior Secretary Carver, dated September 21, 1961. Ibid.

25. Letter of Assistant Interior Secretary Carver to Assistant Defense Secretary Morris, dated November 2, 1961. Ibid.

26. A copy of the Collier lease was provided the author by the Interior Department. It is designated OCS-P-001. The referred to provision is found in section 3(e) 2.

27. Letter of Morris to Secretary Udall, dated May 2, 1962. In the files of the Interior Department, Federal Records Center. Access number 69 A 2058 box 90.

28. Letter of Assistant Interior Secretary Carver to Assistant Defense Secretary Morris, dated June 1, 1962. Ibid.

29. House Report No. 2078, "Ocean Sciences and National Security," op. cit., pp. 5, 17-22, 173.

30. House Appropriations Committee, Subcommittee on the Department of Defense, "Department of Defense Appropriations for 1960," 88th Congress, 1st session, part 1, pp. 757-758.

31. On the Caesar system see Edward Duncan Brown, Arms Control in Hydrospace: Legal Aspects (Washington, D.C.: Woodrow Wilson International Center for Scholars, 1971), pp. 11, 13, and 33; Sven Hirdman, "Weapons in the Deep Sea," Environment 13 (1971): 40-41; John W. Finney, "Secret Undersea System Guards U.S. Against Hostile Submarines," New York Times, September 14, 1965, p. 3.

32. Finney, ibid.

33. Ibid.; and New York Times, September 13, 1965, p. 69.

34. These comments were made by Dr. Craven at the Conference on Law, Organization, and Security in the Use of the Ocean, March 17-18, 1967 at Ohio State University.

35. Ibid.

36. This point was made by an official of the Navy Department during an interview conducted at the Pentagon. The U.S. President's Science Advisory Committee, Panel on Oceanography, Effective Use of the Sea (Washington, D.C.: Government Printing Office 1966), cryptically stated that "In terms of national security we feel high priority should be given to studies of the benthic boundary, since weapon systems of the future may be deployed on the ocean floor . . ." p. 102.

37. Report of the Panel on Oceanography, ibid., p. 33. The Navy was experimenting with small sea-bottom crawling vehicles at least as early as 1960. These devices were known as remote underwater manipulators and were apparently designed for exploration work and possibly for the planting of sonar equipment or mine clearing. See the New York Times, May 22, 1960, p. 32 and May 22, 1961, p. 61.

38. Seymour M. Hersch, "An Arms Race on the Seabed?" War/Peace Report, August-September 1968, p. 21.

39. Hirdman, "Weapons in the Deep Sea," op. cit., p. 36. This was confirmed in an interview with an official of the Navy.

40. Robert A. Frosch, "Military Uses of the Ocean," in Second Conference on Law, Organization and Security in the Use of the Ocean (Ohio State University, 1967).

41. William A. Nierenberg, "Military Oceans," in Nigel Calder, Unless Peace Comes (London: Penguin, 1968), p. 117.

42. The device was designed by Buckminster Fuller and is reported in the New York Times, March 16, 1963, p. 9.

43. Report of the Commission on Marine Science, Engineering and Resources, Our Nation and the Sea (Washington: Government Printing Office, 1969), pp. 5-23.

44. "Seabed Potential for Arms Studied," New York Times, October 10, 1969, p. 7. This information was confirmed in an interview with an official of the Defense Department.

45. Senate Commerce Committee, "Fishing in U.S. Territorial Waters," 88th Congress, 1st session, p. 88. For an impressionistic view of the problems of the American fisherman see E. J. Kahn, Jr., "The Catch in the Hatch Isn't What it Used to Be," New York Times Magazine, August 29, 1971, p. 12.

46. The statistics and a general review of the problems of the American fishing industry are contained in an article by then Senator Benjamin Smith of Massachusetts, "Our Fishing Fleet Comes in Fifth," New York Times Magazine, October 7, 1962, p. 37.

47. D. M. Johnston, The International Law of Fisheries, op. cit., p. 19.

48. William Butler, The Soviet Union and the Law of the Sea, op. cit., p. 189.

49. S. 1988, 88th Congress, 1st session. See also Senator Bartlett's speech introducing this bill in the Congressional Record of August 6, 1963, pp. 14139-14141.

50. House Committee on Merchant Marine and Fisheries, "Fishing in U.S. Territorial Waters," 88th Congress, 2d session, pp. 69-70.

51. Senate Commerce Committee, "Fishing in U.S. Territorial Waters," op. cit., pp. 81-82.

52. House Committee on Merchant Marine and Fisheries, "Fishing in U.S. Territorial Waters," op. cit., pp. 74-75.

53. Ibid., pp. 12-14.

54. 78 Stat. 194.

55. The first list of living species officially declared to be resources of the continental shelf is found in Federal Register 33 (November 2, 1968): 16114. This list includes king crab, tanner crab, and certain species of mollusks and sponges. A notice in the Federal Register 36 (June 23, 1971): 11923 added a few other species such as dungeness crab and the surf clam to this list.

56. Senate Commerce Committee, "Fishing in U.S. Territorial Waters," op. cit., p. 42.

57. Ibid., p. 77. A study produced by the Bureau of Commercial Fisheries in June of 1967, "Oceanography on the Continental Shelf,"

observed that "The shelf and its overlying waters are now the source of 85 percent by volume and 80 percent by value of the fish and shellfish harvested annually by the United States commercial fishing industry." In the files of the Bureau of Commercial Fisheries, Interior Department, Federal Records Center, Suitland, Maryland. Access number 70 A 5547 box 2.

58. Senate Commerce Committee, "Fishing in U.S. Territorial Waters," op. cit., pp. 122-125.

59. 80 Stat. 908.

60. American Journal of International Law 57 (1963); 404.

61. Senate Report No. 1280, 89th Congress, 2d session.

62. House Merchant Marine and Fisheries Committee, Subcommittee on Fisheries and Wildlife Conservation, "Miscellaneous Fisheries Legislation—Fishing Rights," 89th Congress, 2d session, p. 239.

63. See the testimony of Robert E. Giles, General Counsel of the Department of Commerce, ibid., p. 250 and that of Stanley A. Cain, Assistant Secretary of the Interior, ibid., p. 251.

64. Letter of Douglas MacArthur II, Assistant Secretary of State for Congressional Relations, dated May 23, 1966, ibid., p. 252. See also the testimony of Commander Downs, United States Navy, ibid., p. 251 and that of Rear Admiral Wilfred Hearn, Judge Advocate General, United States Navy, ibid., p. 342.

65. Testimony of Commander Downs, ibid.

66. Congressional Record, May 15, 1963, p. 8656.

67. This information was provided the author by the National Oceanic and Atmospheric Agency, Department of Commerce.

68. Clarence Idyll, "The Crab That Shakes Hands," National Geographic 139 (1971): 254-271.

69. This information is derived from an interview with an official of the National Oceanic and Atmospheric Agency in Washington.

70. Windley, "International Practice Regarding Traditional Privileges of Foreign Fishermen in Zones of Extended Maritime Jurisdiction," American Journal of International Law 63 (1969): 492.

71. Treaties and Other International Acts Series (TIAS) No. 5752.

72. TIAS No. 6217, 6635, 7044, 7541.

73. Windley, "International Practice," op. cit., p. 493.

74. TIAS No. 5688.

75. TIAS No. 6155, 6601, 7527.

CHAPTER

6

THE PARDO INITIATIVE TO THE AMERICAN DRAFT TREATY ON THE INTERNATIONAL SEABED AREA

The early 1960s was a relatively quiet period in the development of American thought with respect to seabed matters. The Navy proceeded with its seabed related activities and the Interior Department went ahead with its policy of leasing offshore land in accordance with its interpretation of the Continental Shelf Convention. If doubts were developing in the minds of some Navy or Defense Department officials over possible long-term effects of the policy of the Interior Department in leasing offshore lands farther and farther from the coast and in deeper waters, these doubts had not yet crystallized into firm convictions. The calm, however, was soon to be shattered.

In August of 1967 the Maltese Mission to the United Nations proposed for inclusion in the agenda of the coming 22nd session of the General Assembly an unwieldy item entitled.

> Declaration and treaty concerning the reservation exclusively for peaceful purposes of the seabed and of the ocean floor, underlying the seas beyond the limits of present national jurisdiction, and the use of their resources in the interests of mankind.[1]

The Maltese expressed the fear that advancing technology and existing international law as applied to offshore areas would lead to competition between states for the appropriation and use of more distant offshore areas. This, in turn, it was believed, would be likely to result in the progressive militarization of such areas and to greater levels of international tension. Furthermore, the rewards of offshore exploitation would be limited to only a few states; the potential for benefit to less developed states would be forever lost. In accordance with these views, the delegation of Malta suggested that the time had come to declare the seabed a common heritage of mankind.

In the discussion of this matter Ambassador Pardo of Malta spoke in detail of the need for international action on the seabed. Remarking on the great wealth to be found there and the fact that three-quarters of the earth's land area is covered by oceans, Pardo observed that current international law encouraged the appropriation of this vast area by those with the technical competence to exploit it. He noted that states were leasing offshore land for exploitation at points increasingly distant from shore; in this regard he directed attention specifically toward the United States Department of the Interior and its leasing policies. To put a halt to this process and to allow all states to benefit from the exploitation of the seabed, the Maltese delegate suggested the creation of an international agency with leasing authority over sea bottom areas beyond the limits of national jurisdiction. It was claimed by Ambassador Pardo that by 1975 this agency could have a gross income of about six billion dollars per year, of which, about five billion dollars, and this was said to be a conservative estimate, would be available for development purposes.[2]

It became immediately apparent that the Pardo speech had impressed the representatives of many states.[3] The possibility of creating a new source of much needed development funding was enthusiastically noted by a number of the less developed states. The position of the United States government on the Maltese proposal, however, was friendly but distinctly cautious. In a statement before the first committee of the General Assembly, Ambassador Goldberg indicated that the United States shared the concerns of the Maltese delegation. He noted that on July 13, 1966 President Johnson had stated that:

> Under no circumstances, we believe, must we ever allow the prospects of rich harvest and mineral wealth to create a new form of colonial competition among maritime nations. We must be careful to avoid a race to grab and to hold the lands under the high seas. We must ensure that the deep seas and the ocean bottoms are, and remain, the legacy of all human beings.[4]

Ambassador Goldberg observed, though, that the matter raised by the representative of Malta was very complex indeed and that any hasty action would be imprudent. He suggested that it was too early to make any final decisions on proposals for a comprehensive regime for the deep ocean floor, but noted that the United States would energetically participate in the studies which were needed before such decisions could be made. In terms of immediate action, Ambassador Goldberg called for the establishment of a committee on the oceans to study relevant problems and to make recommendations to the General Assembly.

American caution reflected the uncertainty within the United States government on this issue. By the middle of October 1967, the State Department had prepared a position paper for the United States delegation to the United Nations entitled "The Maltese Proposal Regarding the Deep Ocean Floor."[5] It indicated American concern because the suggestion of Malta might prove to be superficially attractive to many states; the State Department felt that, prior to careful examination, any fundamental decision on the disposition of the resources of the ocean floor by the General Assembly was undesirable. The State Department concluded that Assembly action could best be prevented by developing an attractive and constructive alternative to the Pardo proposal. The alternative decided upon by the Department of State was to provide for the establishment of a General Assembly Committee on the Oceans, to encourage cooperation in the acquisition of knowledge about the ocean and its resources, and to consider general principles to guide states in the exploration and exploitation of the deep ocean floor.[6]

With respect to the establishment of a General Assembly ocean's committee, the State Department had to overcome strong Interior Department opposition. The Interior Department saw in the creation of such a committee "considerable risks and disadvantages for its interests, with few advantages to compensate."[7] The general position of that department was that from both departmental and national points of view, it would be better to deal with seabed matters in specialized agencies rather than in the General Assembly. It was noted that such agencies had the capability to resolve many technical and scientific issues to American advantage with minimum introduction of "extraneous issues." The point was stressed that the impact of United States scientific and technological superiority was greatest in these agencies. More specifically, Interior Department disdain for the handling of seabed matters by the General Assembly, for reasons of departmental interests, is explained by the belief within Interior that:

> For all practical purposes the General Assembly is inaccessible to technical agencies in this Government which could provide the raw material for sound decisions in resource matters. It is necessary to look no further than the recent debates in the General Assembly on the Malta proposal to understand the difficulty. Once the debate began on a U.S. proposal which had been formulated in the most general terms, the pace of the debate and the machinery in this Government precluded any input from this Department in the formulation of subsequent decisions regarding the refinement of the proposal. There

was no opportunity for this Department to examine proposed refinements, to evaluate their implications or to give advice to influence the decision.[8]

Thus, the Interior Department was fearful that its policy input within the government would be minimized by General Assembly consideration of seabed matters. Influence on questions before the General Assembly seemed to gravitate, apparently whatever their substance, toward the State Department and away from "technical agencies" within the government, such as the Department of the Interior. Whatever their merits, the views of Interior did not prevail. The Department of State insisted that an oceans committee was needed and asserted that United Nations practice did not justify the fear that "uniformed diplomatic representatives would attempt to intervene in any substantive way in the work of scientific, technical or legal bodies."[9]

In formulating its response to the Maltese proposal and developing specific alternatives, the State Department was sensitive to other domestic political pressures. It was observed that adequate consultation with Congress and private industry would be required prior to development of principles on the seabed. Meanwhile, the American delegation to the United Nations was authorized to propose a United Nations Decade of Exploration of the Ocean Floor if, in its view, the achievement of a satisfactory alternative to the Maltese proposal required a specific program concept.[10] This option proved unnecessary at the 22nd General Assembly as the assembly was content to establish a 35-member committee to study the peaceful uses of the seabed beyond national jurisdiction and the uses of the resources found there in the interests of mankind.[11]

DEVELOPING CONGRESSIONAL CONCERN

In December of 1967, the House Committee on Foreign Affairs reported on the "United Nations and the Issue of Deep Ocean Resources."[12] The study demonstrated that the State Department had made few, if any, definite decisions as to American policy in this area. At least part of the explanation for State Department timidity is attributed to congressional concern with a possible "give away" of resources considered by some to be in the possession or potential possession of the United States. Representative Rogers of Florida, for example, displayed distrust of the State Department in this regard; he urged Congress to keep close watch upon the actions of the executive branch and to exercise strong congressional oversight on this issue. Rogers claimed that under the Continental Shelf Convention, American claims could extend to mid-ocean if exploitable.[13] Soon after Malta had proposed the seabed item to the General Assembly for its consideration, this type of "nationalist" voice became evident in Congress. In September of 1967, Congressman Pelly of Washington declared that "Committing the United States to turning all the world's deep sea mineral resources over to the United Nations, as the administration apparently plans to do, would constitute the biggest giveaway in the history of America."[14] The administration contacted some members of Congress to assure them that it did not support the full intent of the Maltese proposal or that of Senator Frank Church of Idaho.[15]

Senator Church's plan involved granting title to mineral resources beyond the continental shelf to the United Nations. According to the senator, the adoption of his proposal would remove a future cause of international friction and would also endow the United Nations with a source of substantial revenue.[16] In fact, the idea that the United Nations should hold title to deep sea submerged land was suggested as early as 1962 by New York Representative Benjamin Rosenthal, who noted that United Nations sovereignty in this area would enhance the prestige of the international organization and would result in needed revenue.[17] The belief that a plan such as that of Senator Church or that of Malta might lead to a financially independent or a stronger United Nations provided in itself a motive for opposition to plans of this type. One might, for example, heed the comment of Representative Gross who stated, "I don't think the United Nations stands for much of anything. It never has and probably never will and that is one of the reasons why I don't want to see any authority in this matter vested in the United Nations."[18]

Senator Cotton of New Hampshire, observing that "strong efforts" were being exerted to place the United Nations in immediate control of seabed resources beyond the continental shelf, introduced a resolution that would declare it the sense of the Congress that any action to grant control of deep seabed resources to an international organization would be premature and ill advised.[19] The administration successfully delayed congressional action of this type by indicating that it had no present intention of supporting an effort to vest jurisdiction of the deep ocean floor in an international body.[20] The point was made that the development of a legal regime for submarine areas beyond the continental shelf would not quickly occur; thus, there was no need for congressional action. Senator Cotton, however, claimed that the purpose of his resolution was really to insure a role for Congress in policy formulation on the future development of the ocean bottom.[21] He was reflecting in fact, the growing and more general concern with the institutional position of Congress vis-à-vis the executive branch of government. This evolving attitude was seen in the later remark of Senator Gore, who observed that many people seem to feel that the constitutional responsibility of the Senate in connection with advice and consent is exclusively one of consent.[22] During hearings before the House Foreign Affairs Committee, Congressman Gross asked a State Department witness if Congress would be consulted by the executive branch before definite proposals were made or only after they were made. Without waiting for a response to his question, Gross commented that "we have had some very unpleasant experiences with people in the State Department in taking us into their confidence after the fact."[23] Thus the executive branch was made aware of congressional interest and concern in American seabed policy.

MILITARY INTERESTS AND THE SEABED

For the United States Navy 1967-1969 was crucial period in the development of its attitudes relative to the continental shelf. As noted earlier the Navy, during the early 1960s, was torn between conflicting defense considerations involving, on the one hand, antisubmarine warfare systems and, on the other hand, the possible development of strategic weapons systems to be placed in seabed areas. It was during the 1967-69 period that Defense Department and Navy officials concluded that bottom based strategic weapons systems and submarine bases were not practical. Having eliminated the need for a wide nationally controlled continental shelf for use by weapons systems of these types, it was now possible to concentrate more clearly on the task of creating an effective antisubmarine warfare effort with all that this would imply. One such implication was the need to limit submerged land areas under coastal state control.

The important role of seabed areas in ASW was underlined by the United States position at discussions before the Eighteen Nation Disarmament Conference in 1969. On March 19 of that year the Soviet Union introduced before the ENDC a draft treaty prohibiting all military uses of the seabed and ocean floor beyond a 12-mile limit from the coast.[24] While the United States favored certain restrictions on the military uses of the seabed, the point was emphasized that the United States would not have much interest in a treaty which would eliminate the possibility of all military uses of the seabed.[25] On May 22, 1969, the United States presented a draft treaty to the ENDC that would forbid the emplacement of fixed nuclear weapons or other weapons of mass destruction on the seabed or ocean floor beyond three miles from the coast of any state.[26] The three-mile limit was employed because the United States wanted the prohibition to apply to the maximum practicable area of the seabed. The prohibition contained in the American draft was to apply only to fixed installations and not to mobile systems such as creeping bottom crawlers. When questioned by Senator Case as to why mobile weapons systems were not banned, William Hancock, General Counsel of the Arms Control and Disarmament Agency, observed that such weapons systems posed complex problems in terms of definition and verification; in order to promote immediate progress the text was limited to fixed weapons systems of mass destruction.[27] Negotiations finally led to the February 1971 signing by the United States and the Soviet Union of the Seabed Arms Control Treaty according to which neither state would place fixed nuclear or nonnuclear weapons in or on the seabed beyond the 12-mile limit.[28] Because of its limited purpose, this treaty did not affect the potential antisubmarine warfare uses of the seabed.

In the 1966-67 period another extremely important factor began to enter into Navy thinking with repect to the seabed, namely the phenomenon that has been termed "creeping jurisdiction." In April of 1966, Dr. John Craven of the Navy Special Projects Office publicly raised the important question: to what extent would sovereign rights in the seabed confer additional rights in the waters above?[29] The answer to this question would be of major importance to Navy interests in assuring its long valued freedom of movement in ocean areas. One writer suggested that while the rights conceded by the Continental Shelf Convention are only to exclusive exploitation "it is possible and perhaps probable that these claims will encourage further demand to rights over the bottom, depths and surface of the sea, extending perhaps to the airspace above."[30] This concern was apparently not shared by Dr. Craven who, after some thought on this subject, concluded in March of 1967 that the development of law which reduces international jurisdiction and increase municipal jurisdiction is consistent with the needs of the modern world and provides a long-term, stable solution to the matter of ownership of the ocean bottom.[31]

By late 1967, however, a change had occurred in Dr. Craven's thinking. At that time he noted in an article entitled "The Challenge of Ocean Technology to the Law of the Sea" that developing ocean technology points:

> to the need to preserve the full freedom of the seas in the domain in which the strategic and tactical forces of the Navy have been traditionally deployed . . . within the next decades the seabed will increasingly acquire the nature of real property and the law must continue to grow to protect the investors therein. Such growth apparently competes with the trend of free-surface law. This apparent competition requires most careful and cautious attention to questions of jurisdiction in the seabed.[32]

From the Navy point of view, the Continental Shelf Convention had represented part of a functional approach to the development of the law of the sea. According to Captain Brock of the Navy Judge Advocate General's Office there were two distinct jurisdictional systems within the same geographical column. In the seabed international law has provided the regime of the continental shelf, which is similar in certain respects to the system of territorial jurisdiction found on land. Above the continental shelf, in the water and air space, is the regime of the high seas.[33] The viability of this functionally differentiated perspective was now being questioned in Navy circles.

There can be little doubt that the Maltese initiative at the United Nations put further pressure upon the Navy to develop its position on submerged oceanic lands. The United States government as a whole was now being forced to develop a policy position to present to the international community; Navy and Defense Department input would be a major factor in the determination of American policy.

In late 1967 the Navy requested the Center for Naval Analyses to identify and to assess the impact on naval operations, at present and in the future, of various uses of and activities within the ocean.[34] In its study, the Center noted two important trends. One trend was toward the enlargement of what it called "general purpose zones," areas in which the coastal state was believed to possess general or comprehensive control, jurisdiction, sovereignty, or ownership without providing rights for any previous uses of the area or its resources. The other observed trend was the proliferation of what were termed "special purpose zones," areas in which the coastal state claimed only specific rights for special purposes. An important question to Navy minds involved the relationship between the two types of zones: did concessions on special purpose zones lead to demands for more concessions and eventually to the creation of a new general purpose zone? The center study suggested that the two types of claims were, in fact, intimately related. It was pointed out that:

> statesmen are aware that the most valuable fisheries exist in waters above the continental shelf. When state decision-makers also consider that the waters over the continental shelf are often narrow seas which are highways for marine transportation, and are areas in which some types of naval missions are unique, they have further powerful incentives to try to take over the region generally and monopolize the particular rights.[35]

While the Continental Shelf Convention granted rights to the coastal state only for the purposes of exploration and exploitation of the natural resources of the seabed and subsoil, the study concluded that "rights of national prescription concerning particular activities in the seabed are quite likely to be exercised in a fashion that effectively reduces legal access for other operations."[36]

In its evaluation of the situation the Center for Naval Analyses stressed the importance of ASW systems and cautioned against actions that would interfere with the access of the United States Navy to sea bottom areas for the purpose of placing and maintaining such systems.[37] The overall judgment of the study was that Navy interests were best served by allowing only narrow areas of submerged lands to be placed under national jurisdiction. The belief was explicitly expressed, however, that coastal state claims could not be effectively limited unless there was developed a set of arrangements for the deep seabed. It was felt that potential ocean exploiters could not be expected to be satisfied with a narrow area of firm rights if the area beyond national jurisdiction were left to remain a no-man's-land. In regard to this area, the center study indicated that international devices for allocating and registering commercial rights and claims would be better suited to Navy needs and interests than would schemes for national control of the deep seabed.[38] Thus it was suggested that the Navy had important interests in encouraging the United States government and, through it, the international community to adopt a stable and agreed upon international regime for the lands underlying the deep ocean.

CONCERNS OF THE SCIENTIFIC COMMUNITY

Severe criticism of the present legal regime of ocean space has come from members of the American scientific community. It is claimed that oceanographic research has been hampered, and in some cases prevented, by coastal state action in accordance with present international law. Professor Milner Schaefer, who headed the committee on oceanography of the National Academy of Sciences, claimed that the Continental Shelf Convention was "thoroughly unsatisfactory."[39] First, the point was made that despite the injunction of Article 5 (1) that "the exploration of the continental shelf and the exploration of its natural resources must not . . . result in any interference with fundamental scientific oceanographic or other scientific research carried out with the intention of open publication," fundamental oceanographic research has, in fact, been limited. Brazil, for example, requires details on the institutions and persons sponsoring and conducting research over its continental shelves, proposed routes and working sites and duration of activities, plans and objectives of the research, description of all the equipment to be used, dates of port calls before, during, and after the stay in Brazilian waters, and photographs and technical, scientific and structural specifications of ships or aircraft that are to be used. Moreover, appropriate clearance requests must be filed 180 days prior to the cruise.[40] According to Dr. John Knauss of the University of Rhode Island, the result of these requirements is to effectively prevent scientific research off the Brazilian coast.[41]

Second, Professor Schaefer was critical of the exploitability clause of the Continental Shelf Convention, since it signified an indefinite outer limit to the shelf and thus it remained uncertain as to how far off its coast a state could exercise jurisdiction.

Third, the Continental Shelf Convention left it unclear as to what kinds of research were subject to control by coastal states. This remains a matter of interpretation for each state. On this point the State Department has claimed that: "Research on the continental shelf includes the removal of shelf samples (such as by coring and dredging) and living resources which are unable to move at the harvestable stage except in constant physical contact with the shelf. Research on waters above the shelf or on swimming creatures is not affected." Additionally, in the State Department view, "measurements of magnetic fields of gravity, or the taking of acoustic subbottom reflection measurements, or water samples, would not be considered to be shelf research."[42] This however, is an American interpretation and is not universally accepted. Dr. John Knauss, in testimony before the Senate Foreign Relations Committee, noted that on a recent expedition the research vessel Trident explored continental shelf areas by towing a magnetometer behind it and bouncing high energy sound pulses off the shelf bottom from high pressure air guns also towed behind the vessel. At no time did any of the instruments make physical contact with the shelf in the sense of taking core samples of the bottom. Did this constitute "exploring" the continental shelf within the meaning of that term as employed in the Continental Shelf Convention? According to Dr. Knauss it did not, since there was no physical contact with the seabed. Nevertheless, he pointed out, this position was not held by all states and in order to avoid possible difficulty the Trident obtained coastal state permission anyway.[43]

William Sullivan of the Office of the Special Assistant for Fisheries and Wildlife of the State Department has also observed that freedom of scientific research in ocean areas has been deteriorating because of restraints imposed by coastal states. In 1969 he predicted that the already poor situation would in fact worsen. Sullivan noted that the Continental Shelf Convention differentiated between "exploration," meaning commercial prospecting, and "scientific research," but failed to take into account that the procedures used for these differing purposes were virtually identical. In fact coastal states had increasingly applied the more restrictive controls over commercial prospecting to scientific research with detrimental consequences to the latter.[44]

As a possible solution to such problems, Professor Schaefer returned to the earlier position of the National Academy of Sciences and suggested, in 1969, that the International Council of Scientific Unions or some other suitable international body be designated to certify bona fide scientific research agencies and/or expeditions. He urged that approval of this international body should automatically result in the coastal state's granting of permission to carry out research in the area of its continental shelf.[45]

In June of 1970, the National Academy of Sciences recommended that the United States government consider opening shelf areas subject to American jurisdiction to scientific research by foreign scientists as a means of encouraging other states to ease their own restrictions. The National Academy of Sciences claimed that bold new initiatives on the part of the United States were required to reverse the undesirable trend of increasing restraints on scientific research.[46]

THE STRATTON COMMISSION REPORT

In 1966 Congress passed the Marine Resources and Engineering Development Act, which called upon the President to develop a comprehensive and long range program in marine science.[47] To this end the President was to be assisted by a National Council on Marine Resources and Engineering Development[48] and an independent advisory Commission on Marine Science, Engineering, and Resources.[49] The Marine Council Science was to assist the President in planning and reviewing federal marine science activities. The advisory commission, which became known as the Stratton Commission after its chairman, Dr. Julius Stratton, was to submit to the President and to Congress a report on American marine science activities and recommendations as to future programs and the need for governmental reorganization in the field of marine sciences. On January 11, 1969 the Stratton Commission presented its report.[50] Since the report became a focal point in the discussion and planning of American policy with respect to ocean areas an examination of the report is necessary.

American ocean policy was viewed by the Stratton Commission to be of very great significance; decisions in this area, it was said, would affect the security, economy, and environment of the United States. It was noted that, like the oceans themselves, the interests of the United States in ocean space were vast and complex. What was required was a plan for national action that would accommodate the variety of differing interests. It was strongly suggested that the United States seize the initiative and propose a new international framework for the exploration and exploitation of seabed mineral resources and for scientific research. The Stratton Commission, recognizing the complexity of problems arising from the uses of ocean space, did not recommend a single framework for the management of all the uses of the ocean since, in its view, this was neither feasible nor desirable in the immediate future. The varied uses of the sea presented different problems which called for different solutions.[51]

As to the existing framework of law for offshore mineral exploitation, the Stratton Commission noted that definite uncertainties were apparent relative to the extent of the continental shelf. It seemed obvious that private enterprise would be deterred from exploiting the mineral resources of the seabed under the high seas unless it had firm assurances of exclusive access to such resources. But how far offshore could a state legitimately claim exclusive rights under the exploitability clause of the Continental Shelf Convention? This was a question of very great importance to the American petroleum industry. On July 9, 1968, a committee on the National Petroleum Council issued, on an "urgency" basis, an interim report, which maintained that the exploitability criterion gives coastal states sovereign rights over the natural resources of the geological continental shelves, continental borderlands, continental slopes, and at least the landward portions of the geological continental rise.[52] This point of view was specifically rejected by the Stratton Commission as being unwarranted either by the language of the Continental Shelf Convention or by its history.[53] More significantly, the Stratton Commission took the position that the NPC claim was not consistent with the best interests of the United States since it would create the danger that some coastal states, not possessing important mineral deposits on or under their continental slopes or rises, would feel justified in claiming exclusive access to the superjacent waters, the living resources in them, and the air space above these waters. Such developments were seen as being contrary to American national security, which would best be served by "the narrowest possible definition of the continental shelf for purposes of mineral resource development;" clear concern was thus shown for the phenomenon of "creeping jurisdiction."[54]

The Stratton Commission, consequently, recommended that the United States government attempt to achieve international agreement on a redefinition of the continental shelf for the purposes of the Continental Shelf Convention. It suggested that the outer limit of the shelf be fixed at the 200-meter isobath or at a distance 50 miles from the baseline used to measure the extent of the territorial sea, whichever alternative gave the coastal state the greater area for the purposes of the convention.[55] In the commission's view this would result in a "narrow" shelf with precise outer limits and would thus serve the interests of the United States.

Another important conclusion that the commission arrived at was that the question of fixing the outer limits of the continental shelf was inseparable from that of the framework applicable beyond these limits. It was believed that if coastal states were satisified with the legal-political regime for the exploitation of the mineral resources beyond the continental shelves, they might readily accept the narrow redefinition of the term continental shelf. As a result of this view, the commission urged the United States government to take the lead in proposing a new legal regime for deep ocean space that would encompass the following provisions.

First, there should be an international authority to register national claims beyond the redefined shelf. Registration with the authority would confer exclusive exploitation rights to the area. It was suggested that the authority be part of the United Nations system but should be autonomous in the manner of the World Bank. Additionally, some means of weighted voting based, in large part, on technological capacity was foreseen.[56]

Second, there was to be created an international fund supported by a portion of the value of production derived from claims registered by the international authority. This fund was to be used to finance marine scientific research and to aid developing states through the World Bank, the United Nations Development Fund, and other international agencies. It was specifically recommended that the proceeds from the collected funds should not be employed for general purposes of the United Nations. Clearly, the Stratton Commission did not want discussion and debate on oceans policy to become bogged down because of differences as to the desirability of a financially independent United Nations; at the same time there was a recognized need for some sort of international program for fund use, which would be internationally attractive.

Third, certain responsibilities were placed on the registering nations. Each nation making a claim would have to see to it that the business entity on whose behalf the claim was made complied with the conditions of the international registry authority and that specified fees and payments were submitted. While each state would police the activities occurring under its registered claims, the authority would be empowered to cancel a claim if the registering nation failed to discharge its obligations. The international authority, further, was to be allowed to inspect all installations and equipment used in operation under a registered claim.[57]

Fourth, the international registry authority was to settle whatever disputes arose under the recommended framework.[58]

Fifth, it was suggested that there be created an offshore intermediate zone, which would extend to the 2,500-meter isobath or to a distance of 100 miles from the baseline for measuring each nation's territorial sea, whichever alternative gave the coastal state the greater area.[59] In it, only the coastal state and those who were licensed by it could explore and exploit mineral resources. The creation of this zone was seen as providing a reasonable compromise between the view that the definition of the continental shelf should include the intermediate zone area and the position that the intermediate zone area should be treated in all respects like the areas of the deep seas beyond it. The intermediate zone was a key concept for the Stratton Commission since it was believed that it would be widely accepted and would serve to prevent claims of exclusive access to the area for purposes other than general resource development. A portion of the value of the production from this zone was to be paid into the international fund.[60]

It was understood by the members of the Stratton Commission that it would take years before a new framework such as it had suggested would be created; the commission, therefore, offered a course of interim action. It recommended that the United States propose the principle that no nation, in the interim period, should claim or exercise sovereignty or sovereign rights over any part of the seabed or subsoil beyond the 200-meter isobath. The possibility of exploitation beyond the 200-meter depth was not foreclosed during this interim period, however, since the commission suggested that the United States continue to authorize exploration and exploitation of mineral resources beyond this limit. Authorizations to operate beyond 200 meters, though, were to state explicitly that such operations would be subject to whatever international regime was eventually agreed upon. In an attempt to provide for what was regarded as the necessary security of investment if private enterprise were to operate beyond the 200-meter limit, the commission called upon Congress to enact legislation providing for compensation for losses due to an international framework that placed areas in which seabed exploitation was occurring beyond the outer limits of the continental shelf.[61]

The Stratton Commission saw a need for immediate action. As the United Nations had become involved in the issue of the seabed it was now imperative that the United States develop a carefully formulated position so that it could take the initiative in shaping a framework for the oceans that would properly reflect significant American interests.

Fisheries

With respect to fisheries, the International Panel of the Stratton Commission specifically evaluated and rejected two alternatives. It rejected the suggestion that each coastal state be given permanent and exclusive access to the living resources in the waters superjacent to its continental shelf. This proposal, it was believed, would encourage states to declare their sovereignty over this area with resultant interference in other uses of the sea such as navigation. Moreover, this alternative would not assure the most productive use of fishery resources in the expanded exclusive fishing zones since the coastal state might not have the ability to exploit these resources fully. The commission also turned down a proposal that would have given to the United Nations, in the name of the international community, title to the living resources of the high seas beyond the 12-mile limit. To entrust any agency with a monopoly over the exploitation of the living resources of the high seas was to take very great risks unwarranted by the needs of the existing situation.[62]

The panel concluded that a better approach to the problem of the living resources of the oceans was through the fixing of national catch quotas in certain fishing areas. A quota had the advantage that it could be designed to maintain the maximum sustainable fishery yield.[63] But this alternative, too, obviously presented certain problems. Aside from the matter of getting the states already fishing in a given area to agree as to what catch they should accept, a serious obstacle would be presented by prospective new entrants to established fisheries.[64] It was clear that fishery problems were complex and that a generally acceptable agreement would not be easy to achieve.

Scientific Research

The commission noted that the existing international law of ocean areas did not facilitate scientific investigation. It was noted that there were many ambiguous provisions in the Continental Shelf Convention, which could be interpreted by coastal states so as to increase the restrictions on scientific research. To deal with the obstacles to legitimate scientific inquiry, the commission urged the United States to join with other states to establish the principle of maximum freedom for scientific research. What was required, according to the commission, was a new international regime, embodied in treaty law.

Rather sweeping suggestions were made by the Stratton Commission in the principles proposed for this new convention on scientific research. Research in a state's territorial waters or on its continental shelf could be conducted without its consent if the coastal state were informed of such research and if the coastal state were given the opportunity to participate in the research, the results of which were to be published. Fisheries research, which could involve the capture of specimens, could be conducted in the exclusive fisheries zone of any coastal state under the same conditions. Submersible research vessels could be employed for scientific research in the territorial seas of coastal states after proper advance notification had been given. Finally, research buoys could be placed in a state's territorial seas or in the waters over its continental shelf. Such buoys were to be protected against unwarranted interference from any source.[65]

One might well expect that security conscious state officials would be extremely wary of these recommendations. To lessen these suspicions, the commission noted that since the coastal state could always choose to participate in the research off its coast, there would be a built-in safeguard assuring that scientific inquiry would not become a cover for espionage.[66]

Until a new framework for scientific research could be achieved, the commission recommended that the United States seek bilateral and regional agreements containing the provisions suggested for the new convention. Additionally, the United States was urged to take other initiatives to encourage freedom of scientific research and international scientific cooperation. Among the suggestions made was that the United States consent to the conduct of any proposed foreign scientific research certified by the Intergovernmental Oceanographic Commission as meeting the requirements of the Continental Shelf Convention.[67]

THE PETROLEUM INDUSTRY AND THE SEABED

Soon after the publication of the Stratton Commission Report, the National Petroleum Council made public a report of its own entitled Petroleum Resources Under the Ocean Floor.[68] This study was undertaken by the NPC in pursuance of a request by the Department of the Interior. The report drew a number of conclusions. The NPC estimated that domestic demand for crude oil would rise by more than 50 percent between 1965 and 1980 and stressed that a healthy and expanding domestic petroleum industry was essential to the security of the United States and to that of the free world.[69] It observed that a major promised source of United States petroleum supplies for the future lay in the resources "of the entire submerged portion of the continent extending from the coasts of the United States and down to the abyssal ocean floor;" in addition, prospects of significant finds of petroleum under the deep oceans could be neither confirmed nor ruled out as possibilities.[70] In relation to deeper water operations, the progress of the oil industry was said to be limited primarily by economic attractiveness rather than by technological capability. It was claimed that within five years technology would allow exploitation in water depths up to 1,500 feet and within ten years up to depths of 6,000 feet.[71] According to the NPC, the United States and other coastal states possessed exclusive sovereign rights to natural resources underlying the seabed of the entire submerged portion of the continent adjacent to their coasts. The NPC strongly stressed the point that coastal jurisdiction over subsea mineral resources did not affect the high seas status of superjacent waters; article three of the Continental Shelf Convention was said to make this quite clear.[72] As a consequence of these perceptions, the NPC urged the United States to assert promptly and clearly its exclusive jurisdiction over the mineral resources of the entire submerged portion of the continent off its shores.[73] It was suggested that not to take this course of action, in the light of our present and future energy needs, would risk the vital national interests and the security of the United States and also that of American allies. Additionally, the NPC pointed out that the federal treasury had an important stake in the geographical extent of American jurisdiction over petroleum resources since it derived very substantial revenues from oil exploitation on federally owned land.[74] Finally, it was claimed that existing principles of international law were adequate to govern petroleum exploration and exploitation for some time to come; the state of knowledge about the deep ocean areas was such that it would be premature to plan now for its development.[75] In its view, the NPC preference as to the outer limit of the continental shelf could well be supported by the wording and intention of the Continental Shelf Convention.[76]

After the publication of the NPC study, the American Petroleum Institute issued a statement of policy on the matter of jurisdiction over the natural resources of the ocean floor that mirrored the concerns of the NPC and supported NPC conclusions.[77] Backing for the NPC-API view was voiced in the editorial policy of the Oil and Gas Journal, which attacked what it termed the "giveaway crowd," which would surrender full and exclusive sovereignty to part of the continental margin.[78]

Senator Claiborne Pell, in his critical analysis of the preferences of the petroleum industry, alleges that the industry position had been shaped, to a great extent, by the oil import quota and the benefits that were derived from foreign tax credits. In reference to the import quota program, the closer the offshore boundary is drawn to shore the greater the amount of offshore oil that would become subject to the quota. As to foreign tax credits, the oil companies must be doing business with a foreign government in order to claim such credits. Senator Pell has suggested that the petroleum companies may have had doubts about whether they would continue to receive tax credits and expense write-off benefits if their activities were being conducted under the administration of an international licensing authority.[79] That such considerations have been factors in determining the industry's position on seabed matters has been denied vehemently by petroleum spokesmen.[80]

INTEREST ARTICULATION BEFORE CONGRESS

With the introduction before the United Nations of the issue of the seabed, and with the growing awareness that sharp differences with respect to the subject existed inside and outside of the United States government, several committees of the Congress began to examine the situation.[81] At the various congressional hearings the divisions of interest were made apparent.

The Defense Department and the Navy emphasized their determined preference for narrow offshore areas under coastal state jurisdiction. In testimony before the Senate Foreign Relations Committee in 1969, Robert Frosch, Assistant Secretary of the Navy, expressed the firm Navy position that "the narrowest definition of the continental shelf would be best . . . both from a national security point of view and in keeping with the freedom of the seas attitude."[82] The concern of the Navy with "creeping jurisdiction" was evident in Frosch's thought as he testified that we must be aware of the problem of "spillover" of jurisdiction from the seabed to the waters above. On this point Warren Nutter, Assistant Secretary of Defense, explained that "the overriding concern of the Defense Department is that the eventual legal regime for the seabeds have no effect on the traditional freedom of the seas. All too often, we have seen legal regimes applicable to one area spill over into others."[83]

Testifying before the Senate Foreign Relations Committee, Cecil Olmstead, a representative of the petroleum industry, questioned whether the fears expressed by Defense and Navy officials were warranted. He claimed that under the Continental Shelf Convention the character of the seas above the continental shelf remained unaffected and that traditional uses of the sea, such as navigation and fishing, were left undisturbed.[84] In an appearance before the Senate Commerce Committee, Olmstead maintained that the United States "already has exclusive jurisdiction over the natural resources of the continental mass seaward to where the submerged portion of that mass meets the abyssal ocean floor. We believe this Nation, in concert with other like-minded countries, should declare its rights accordingly."[85] In later testimony before the Senate Interior Committee, he indicated that the major policy consideration in the matter of offshore land is the national security of the United States; with respect to this consideration the petroleum industry had a different emphasis and had reached different conclusions than had the Navy and Defense Departments. Noting that national security is composed of a variety of elements, Olmstead indicated that among the most important was the "economic base" which supported military forces. Petroleum, he pointed out, is a vital ingredient to national security since the military forces of the United States are exceedingly dependent upon it. The increasing need for petroleum in the United States taken together with the wish to limit American dependence on foreign oil dictated the need to support a wide shelf alternative. According to Olmstead "no known advantage would accrue to the United States in terms of either long- or short-range economic and national security interests in opting for a narrow shelf."[86] Among disadvantages said to be inherent in a narrow shelf concept was that the United States Treasury would be deprived of anticipated revenues from minerals in offshore areas beyond the narrowly defined continental shelf. Noting that up to the end of 1968 the Treasury had received about 4 billion dollars from offshore petroleum bonuses on lease sales and royalties on production, Olmstead suggested that revenues of this magnitude and more are likely from the area lying between a narrow and a wide shelf.[87]

Supporting the position of the petroleum industry before the Senate Interior Committee was Northcutt Ely, a member of the Council of the Section of Natural Resources of the American Bar Association, who also suggested that a wide shelf area under national jurisdiction would be very beneficial to the United States Treasury. In addition, Ely stressed that the issue of control was an essential consideration; it should be a matter of Congress to decide, rather than some international body, how and when resources off American coasts were to be exploited.[88] Further supporting the preferences of the petroleum industry was the testimony of Dr. Miller Spangler of the Center for Techno-Economic Studies of the National Planning Association. He noted that there was increasing expectation that the continental rise at depths of 2,000-12,000 feet would prove to have sizeable oil deposits. Accordingly, he asserted that the United States would serve its best interests by retention of the existing Continental Shelf Convention, which was said to permit national exploitation of the resources of the continental shelf, slope, and rise. Aside from the direct benefits from marine mining, it was suggested that the United States could ease its balance-of-payments problems by mining a wider shelf as opposed to importing needed minerals. Moreover, American dependence upon politically unstable parts of the world for strategic materials affecting national security would be reduced.[89]

In testimony before the Senate Commerce Committee in 1969, officials of the Interior Department presented a point of view similar to that of the NPC and the API.[90] William Pecora, Director of the Geological Survey, observed that the Interior Department considered the limit of the continental shelf to be at an average of 2,500 meters under the terms of the definition found in the Continental Shelf Convention. In some places this would result in American jurisdiction as far from shore as 250 miles.[91] Summing up the Interior Department's preference as to what American policy should be, Russell

Train, Interior Undersecretary, asserted than the United States "should claim jurisdiction over the resources of the entire Continental Shelf right out to the edge of the deep ocean bottom."[92]

Senator Hollings, having listened to the testimony of Interior Department witnesses, pointed out that their position was considerably different from that of the Defense Department. The senator was told that representatives of the Department of State, Defense, and Interior were attempting to resolve their differences; Interior Undersecretary Train took the opportunity to suggest that the fear of the Defense Department of "creeping jurisdiction" was uncalled for. In his view there was no legal basis for jurisdictional spillover and this was made abundantly clear by the Continental Shelf Convention.[93]

While the positions of the Interior and Defense Department were established the views of the State Department remained unclear. In testimony before the Senate Foreign Relations Committee in July of 1969, U. Alexis Johnson stated: "Frankly, Mr. Chairman, the question of the boundaries, the question of the international regime, are questions the answers to which are not yet clear to me, nor am I clear if I may say, both personally and officially, as to where the U.S. interests lie best in this."[94] Apparently, the State Department was not very strongly convinced by the theory of "creeping jurisdiction." In hearings before the Commerce Committee Senator Hollings asked a State Department assistant legal adviser if a boundary could be set just for the seabed with the waters above remaining unaffected. The response was: "Yes sir . . . it has been our approach to keep separate the question of a boundary for coastal state interest in the seabed and coastal state interest in the water column. It has been an approach that has been followed by the United States since at least 1945."[95]

Though the State Department was unsure as to where the boundary of national jurisdiction in offshore areas should be and was not overly worried about creeping jurisdiction, it was growing concerned over the possibility of seabed claims becoming a cause of international conflict. The expectation that great wealth would be found in the seabed, it was feared, could stimulate an escalation of claims and result in serious disputes between states, some of which might involve the United States either directly or indirectly. As the executive department primarily concerned with foreign relations, it is not surprising that the State Department concluded that it was in the long-range interests of the United States to assure that questions involving matters such as boundaries and the regime for the seabed were resolved by widely accepted international agreement.[96] The Defense Department strongly agreed that these matters had to be settled by multilateral treaty since the alternative possibility of developing the law of the sea by unilateral claims, it was felt, would lead to consequences that would be highly detrimental to American defense interests.[97]

One might well conclude from this that while the Interior and Defense Departments were fighting over substantive differences on the question of where the boundary of the continental shelf should be, the State Department was chiefly concerned with achieving an internationally acceptable boundary that would serve to lessen the possibility of conflict over the seabed.

Among the other interested parties appearing at congressional hearings was Malcolm Wilkey, general counsel to Kennecott Copper, who noted the interest of his industry in the type of regime for deep ocean space, where manganese nodules are found. Concern with the seabed was already developing in the areas beyond the continental margins. Wilkey urged the creation of an international regime that would allow and encourage the exploitation of the deep seabed by private enterprise.[98]

PRESIDENTIAL LEADERSHIP AND THE DRAFT CONVENTION

With the introduction of the subject of the seabed before the United Nations it became increasingly important for the United States to develop some relatively clear position. As seen, sharp and significant differences had emerged inside the United States government with respect to what American policy should be. For various reasons the Interior Department favored a wide continental shelf under coastal state jurisdiction; the Defense Department and the Navy strongly preferred that only very narrow offshore areas be under coastal state control. Since agreement as to policy could not be reached at the interdepartmental level, the decision was pushed up to the level of the President. On May 23, 1970 President Nixon made an important announcement on United States ocean policy.[99]

The President noted that the nations of the world had grown increasingly conscious of the wealth to be exploited from the seabed and the waters above. At the same time the law of the sea lagged behind technical advances and consequently required updating; according to the President, if the law were "not modernized multilaterally, unilateral action and international conflict are inevitable." With this consideration in mind, President Nixon proposed that:

> All nations adopt as soon as possible a treaty under which they would renounce all national claims over the natural resources of the seabed beyond the point where the high seas reach a depth of 200 meters (218.8 yards), and would agree to regard these resources as the common heritage of mankind.

In this statement it can be seen that the position of the Defense Department favoring a narrow shelf option was acted upon by the President. Further, his rejection of the existing law of the sea as "inadequate" conflicted with the explicit views of the NPC and API.

The President went on to suggest that the treaty should establish an international regime for the exploitation of seabed resources beyond the 200-meter limit. According to the President:

> The regime should provide for the collection of substantial mineral royalties to be used for international community purposes, particularly economic assistance to developing countries. It should also establish general rules to prevent unreasonable interference with other uses of the ocean, to protect the ocean from pollution, to assure the integrity of investment necessary for such exploitation and to provide for peaceful and compulsory settlement of disputes.

The first sentence of the above citation obviously is directed to the international audience, assuring it that there is much to be gained by adoption of a treaty of the type proposed by the President. In the second sentence the desire to prevent "unreasonable interference with other uses of the ocean" is addressed to the Navy and the fishing and scientific communities. The reference to "integrity of investment" is clearly aimed at soothing the fears and doubts of the petroleum industry which harbors deep suspicion toward any type of international regime having jurisdiction over seabed areas. In a further effort to win industry support, the President indicated that he would propose a change in United States tax legislation so that Americans who operated in the seabed beyond exclusive American jurisdiction were not discriminated against as compared to Americans undertaking commercial activities within the territory of the United States.

President Nixon, following in part the recommendations of the Stratton Commission, suggested a two zone arrangement beyond the portion of the seabed under exclusive national control. He called for the creation of an international trusteeship zone in the area of the continental margin past the 200-meter depth. In this area the coastal state would act as trustee for the international community and would receive some part of the international revenues derived from the zone. Beyond the area of the continental margins an international regime would be empowered to authorize and to regulate the use of seabed resources. The President's policy statement did not elaborate upon the trusteeship zone nor upon the international machinery to be created but merely noted that the United States would introduce specific proposals before the United Nations Seabed Committee.

Recognizing that negotiations on a seabed treaty would take some time, President Nixon called upon all states to join the United States in an interim policy according to which "all permits for exploration and exploitation of the seabed beyond 200 meters be issued subject to the international regime to be agreed upon." This policy was clearly different from that urged by the United Nations General Assembly in 1969 when it passed the "Moratorium Resolution" over United States opposition. That resolution stated in part that pending the establishment of an international regime for the seabed "States and persons, physical and juridical, are bound to refrain from all activities of exploitation of the resources of the area of the sea-bed and ocean floor, and the subsoil thereof, beyond the limits of national jurisdiction."[100] Within the United States government this resolution was opposed by the Interior Department, which did not desire limitations on its right to lease, and also by the Defense Department. Defense feared that the Moratorium Resolution would, in fact, encourage states to make even wider national claims to offshore areas since the "limits of national jurisdiction" were not defined in that resolution. A Defense Department spokesman observed that in 1970 Brazil, for example, claimed a 200-mile territorial sea after supporting the 1969 resolution.[101] Interior and Defense were thus united in opposition to the Moratorium Resolution but for different reasons.

Four days after the presidential statement, Undersecretary of State Elliot Richardson appeared before the Senate Interior Committee to explain the President's policy statement. An analysis of Richardson's presentation to the committee suggests that all of the following interests would be served by a multilateral treaty embodying the concepts proposed by President Nixon:

1. the defense interest of the United States concerned with free navigation and "creeping jurisdiction";[102]
2. the scientific community which desired maximum freedom for research;[103]
3. the interests of the coastal state in controlling what occurs off its coasts;[104]
4. the interests of the international community in sharing part of the proceeds of the exploitation of the seabed;[105]
5. the investors in seabed exploitation who needed a set of established and stable rules.

In this last regard the point was made that under the President's proposal companies would continue to deal with the coastal state out to the edge of the continental margin because of the trusteeship zone.[106] The petroleum industry was also specifically assured that petroleum taken from the trusteeship area of the United States would not be considered as being imported and thus subject to oil import quotas. John Stevenson, State Department Legal Adviser, did note under questioning, however, that this was a matter upon which Congress would have to legislate.[107] As could be seen then, the President's expressed policy was designed carefully to accommodate different interests in the domestic and international political environments.

On August 3, 1970 the United States introduced before the United Nations Seabed Committee a draft treaty that elaborated upon the basic views expressed by the President in his May 23, 1970 announcement.[108] The proposed treaty called for the recognition of an international seabed area that would encompass all the seabed and subsoil of the high seas beyond the 200-meter isobath adjacent to the coasts of continents and islands. The international seabed area was in turn divided into two parts. The part closer to the coast, extending from the 200-meter depth line to the base of the continental slope, was designated as the International Trusteeship Area.[109] In this portion of the seabed the coastal state was to act as trustee for the international community, exercising only those powers specifically provided for in the treaty. The powers granted the coastal state in this regard included the issuing and suspending of exploration and exploitation licenses, setting work requirements, policing licensees, and enacting regulations as were required.[110] The area seaward of this trusteeship zone was to come under the direct control of a proposed international organization, the International Seabed Resource Authority. In no part of the international seabed area could any state claim or exercise sovereignty or sovereign rights.

The key feature of this plan was the trusteeship zone and the compromises it embodied. Care was given to the concerns of the coastal state; it would determine who would operate in this zone and it would provide the civil and criminal legal framework under which exploitation would occur.[111] At the same time, the international community would benefit from payments derived from this area and made to the International Seabed Resource Authority, which was to use such funds "for the benefit of all mankind, particularly to promote the economic advancement of developing States Parties to this Convention, irrespective of their geographic location."[112] This provision clearly sought to win support for the draft treaty among the less developed states and those states which were landlocked or shelf-locked. Such states were said to have further assurances that they would possess a significant voice in seabed matters because of the manner in which the International Seabed Resource Authority Council was to be constructed. At least 12 members of the council would have to be developing states and at least 2 would have to be landlocked or shelf-locked states out of a total council membership of 24 states. Additionally, the trusteeship zone as envisaged would serve to protect the security concerns of the Navy since navigational rights and overflight would not be affected. Finally, recognition for the plight of the scientific community in seabed matters was also demonstrated. According to article 24 (1) "Each Contracting Party agrees to encourage, and to obviate interference with, scientific research." This provision however, left unanswered the question, as to what, exactly, constituted "interference" with scientific research.

Seemingly, then, the United States had formulated a specific plan that reconciled the multiple uses of ocean space and also protected significant American interests. In actual fact, however, the political battle within the United States over oceans policy was only beginning.

NOTES

1. Note verbale from the Permanent Mission of Malta to the United Nations Secretary-General, dated August 17, 1967, circulated as A/6695.

2. A/C. 1/P.V. 1515, 23-67; A/C. 1/P.V. 1516, 3-7. According to Pardo the $6 billion figure was based upon the supposition that the boundary limiting national seabed jurisdiction would be fixed at either the 200-meter isobath or the 12-mile limit.

3. See, for example, the statements of the representatives of the following states: Norway, A/C. 1/P.V. 1525, 22-25, the Netherlands, ibid., 26-30; Libya, ibid., 36-42; Somalia, ibid., 43-44; Nigeria, A/C. 1/P.V. 1526, 21-22; Ghana, ibid., 36-41; Trinidad and Tobago, ibid., 46-50; Ceylon, ibid., 50-53; Austria, A/C. 1/P.V. 1527, 41-46; Sweden, ibid., 46-56; Tanzania, ibid., 56-66; Colombia, A/C. 1/P.V. 1528, 2-7; Sierra Leone, ibid., 13-16; Italy, ibid., 16-21; Liberia, ibid., 22-28; Turkey, ibid., 43-46; UAR, ibid., 47-56; Afghanistan, ibid., 61-65; Tunisia, A/C.1/P.V. 1529, 11-12; Mexico, ibid,, 28-37; Yugoslavia, ibid., 62-66; Canada, ibid., 67-75; Jamaica, ibid., 76-81; Iran, ibid, 82-87; and India, A/C. 1/P.V. 1530, 2-6.

4. A/C. 1/P.V. 1524. The quoted statement of President Johnson was made at the commissioning of the new research ship, Oceanographer. The full text of his remarks appears in Weekly Compilation of Presidential Documents 2 (1966): 930-931. The statement indicates that high officials were beginning to be concerned with the international ramifications of seabed exploitation.

5. A copy of this position paper, dated October 19, 1967, was found in the files of the Bureau of Commercial Fisheries, Interior Department, Federal Records Center, Suitland, Maryland. Access number 70 A 5547 box 5. See pp. 1-3.

6. Ibid., p. 2.

7. Department of the Interior, "Position Paper on U.S. Response to U.N. Questionnaire on Ocean Resources Research and Utilization." A copy of this paper, dated February 1968, was found in the files of the Bureau of Commercial Fisheries, Federal Records Center, Suitland, Maryland. Access number 70 A 5547 box 5. See pp. 2, 7.

8. Ibid., p. 7.

9. Paper prepared by D. H. Popper, Department of State, "Need for Committee on the Oceans." Dated February 6, 1968, Ibid., p. 5.

10. "The Maltese Proposal Regarding the Deep Ocean Floor," cited in footnote 5. See pp. 5-6.

11. General Assembly Resolution 2340 (XXII), December 18, 1967.

12. House of Representatives, Report No. 999, 90th Congress, 1st session.

13. Ibid., pp. 40, 46-47.

14. Congressional Record, September 14, 1967, p. 25649.

15. Congressional Record, September 14, 1967, p. 25659.

16. Congressional Record, August 25, 1967, pp. 24189-24190.

17. New York Times, July 15, 1962, section 4, p. 8.

18. House Committee on Foreign Affairs, "The United Nations and the Issue of Deep Ocean Resources," 90th Congress, 1st session, p. 100.

19. S.J. Resolution 111, 90th Congress, 1st session.

20. Letter from Assistant Secretary of State William B. Macomber, Jr. to Senator Fulbright, dated November 1, 1967. Senate Foreign Relations Committee, "Governing the Use of Ocean Space," 90th Congress, 1st session, pp. 10-11.

21. Ibid., p. 16.

22. Ibid., p. 17.

23. House Committee on Foreign Affairs, "The United Nations and the Issue of Deep Ocean Resources," op. cit., p. 56.

24. The text of the Soviet proposal appears in Senate Foreign Relations Committee, "Activities of Nations in Ocean Space," 91st Congress, 1st session, pp. 16-17.

25. See the statement of Gerard Smith, Director of the United States Arms Control and Disarmament Agency quoted in the New York Times, March 21, 1969, p. 21. A similar statements was made by William Hancock in testimony before the Senate Foreign Relations Committee, "Activities of Nations in Ocean Space," op. cit., p. 6. The position of the United States as expressed before the Seabed Committee was that the seabed should be used only for peaceful purposes; the American representative suggested that "the test of a 'peaceful' activity being whether it was consistent with the United Nations Charter and other obligations of international law. Since,

however, 'peaceful purposes' would not preclude all military activities, a detailed arms control agreement would have to be evolved to prohibit specific military activities." A/AC. 135/SR. 17, p. 52. The representative of Great Britain was more explicit in explaining his country's desire to maintain the right to certain defensive uses of seabed areas for purposes of submarine surveillance. A/AC. 135/SR. 17, pp. 53-54.

26. The text of the American proposal appears in "Activities in Ocean Space," op. cit., pp. 15-16.

27. Ibid., p. 5.

28. Senate Foreign Relations Committee, "Seabed Arms Control Treaty," 92nd Congress, 2d session, pp. 39-41.

29. John Craven, "Sea Power and the Sea Bed," U.S. Naval Institute Proceedings 92 (1966): 49.

30. L. W. Martin, The Sea in Modern Strategy (New York: Praeger Publishers for the Institute of Strategic Studies, 1967), p. 21.

31. John Craven, "Technology and the Law of the Sea," in Conference on Law, Organization and Security in the Use of the Ocean, 1967, op. cit., 2: 35-36.

32. JAG Journal 22 (1967): 38.

33. John Brock, "Mineral Resources and the Future Development of the International Law of the Sea," JAG Journal, ibid., p. 44.

34. The detailed findings of this study are found in N. V. Breckner, The Navy's Role in the Exploitation of the Ocean (Project Blue Water) Phases I, II, and III (The Center for Naval Analyses, 1968-1970). Phases II and III are classified documents but in 1972 the material contained in all three phases was edited and brought together in one unclassified volume entitled The Navy and the Common Sea (Washington, D.C.: Office of Naval Research, 1972). In a letter to the author from a former official of the Navy Hydrographic Office, it was suggested that the "Blue Water" studies had considerable influence on Navy thinking.

35. The Navy and the Common Sea, ibid., pp. 142-144.

36. Ibid., pp. 142-145, 191-192, 210-211.

37. Ibid., pp. 202, 209.

38. Ibid., pp. 211-212. It was specifically noted that "nationalization schemes for the deep seabed would impinge more on the U.S. Navy in the future than certain international devices for allocating commercial rights and transferring claims."

39. Milner B. Schaefer, "Freedom of Scientific Research and Exploration in the Sea," Stanford Journal of International Studies 4 (1969): 63.

40. The text of the detailed Brazilian regulations appear in "Activities in Ocean Space," op. cit., pp. 115-118.

41. Ibid., p. 113. See also Louise Purrett, "The Politics of Marine Research," Science News 99 (January 2, 1971): 9-11.

42. Milner B. Schaefer, "The Changing Law of the Sea—Effects on Freedom of Scientific Investigation," Proceedings of the Second Annual Conference of the Law of the Sea Institute, op. cit., p. 115.

43. "Activities in Ocean Space," op. cit., p. 107.

44. William Sullivan, "Freedom of Scientific Research in Conflict With the Development of Ocean Resources," p. 1. This paper was presented at a regional meeting of the American Society of International Law in Miami on December 13, 1969.

45. Schaefer, "The Changing Law of the Sea," op. cit., p. 117.

46. National Academy of Sciences Press Release, dated June 11, 1970.

47. Public Law 89-454.

48. This Council was to be chaired by the Vice-President and to be included as members were the Secretary of State, Secretary of the Navy, Secretary of the Interior, Secretary of Commerce, Chairman of the Atomic Energy Commission, Director of the National Science Foundation, Secretary of HEW, and the Secretary of Transportation.

49. On January 9, 1967 President Johnson appointed the following people to the Commission: Dr. Julius Stratton, Ford Foundation (chairman); Dr. David Adams, Department of Commercial and Sport Fisheries, State of North Carolina; Dr. Carl Auerbach, University of Minnesota; Jacob Blaustein, Standard Oil; Professor James Crutchfield, University of Washington; Leon Jaworski, attorney; Dr. John Knauss, University of Rhode Island; John H. Perry, Perry Publications; Taylor Pryor, Sea Life, Inc.; George Reedy, Struthers Research and Development Corp.; Dr. George H. Sullivan, Northrop Space Corp.; Robert Baldwin, Undersecretary of the Navy; Frank Di Luzio, Assistant Secretary of the Interior for Water Pollution Control; Robert White, Environmental Science Services Administration; Senators Norris Cotton and Warren Magnuson; and Congressmen Alton Lennon and Charles Mosher.

50. The lengthy report of the Commission on Marine Science, Engineering and Resources is entitled Our Nation and the Sea: A Plan for National Action (Washington, D.C.: Government Printing Office, 1969). This report is supplemented by three volumes of panel reports. Volume I, Science and Environment; volume II, Industry and Technology; volume III, Marine Resources and Legal-Political Arrangements for Their Development. According to the foreword in the commission report, the panel reports "constitute the primary source material upon which the Commission based its own final conclusions."

51. Stratton Commission Report, op. cit. p. 1-33.

52. National Petroleum Council, Petroleum Resources Under the Ocean Floor: An Interim Report of the National Petroleum Council (Washington, July 9, 1968).

53. Stratton Commission Report, op. cit., p. 4-III-47. See also the study of Professor Louis Henkin, Law for the Sea's Mineral Resources (Washington, D.C.: National Council on Marine Resources and Engineering Development, December 1967), especially pp. 37-38, 99. It is likely that this study influenced the views of the Stratton Commission.

54. Stratton Commission Report, op. cit., p. 4-III-49 to 50. The Commission noted that "The danger that rights of exclusive access for one purpose may expand to claims of territorial sovereignty or exclusive access for all purposes materialized as an unfortunate and undesirable consequence of the Truman Proclamation of 1945."

55. Ibid., pp. 4-III-50 to 51. According to the commission 200 meters is the average depth of the outer edges of the world's geological continental shelves; 50 nautical miles is the average width of the shelves.

56. Ibid., pp. 4-III-57 to 61.

57. Ibid., pp. 4-III-63 to 64.

58. Ibid., pp. 4-III-64 to 66.

59. Ibid., pp. 4-III-68 to 70. The 2,500-meter isobath was chosen because it was said to be the average depth of the bases of the world's geological continental slopes; 100 nautical miles was said to be the average width of the continental shelves and slopes.

60. Ibid.

61. Ibid., pp. 4-III-76 to 77.

62. Marine Resources and Legal-Political Arrangements, op. cit., p. viii-57.

63. Ibid., pp. viii-57 to 59.

64. Ibid., pp. viii-63 to 65.

65. Stratton Commission Report, p. 5-79.

66. Ibid., p. 5-82. Coastal state sensitivity on this point could only have been intensified as a result of the 1968 incident involving the United States vessel Pueblo and North Korea. On the Pueblo affair see the Proceedings of the American Society of International Law, 1969, pp. 2-30.

67. Ibid., p. 5-85.

68. National Petroleum Council (Washington, D.C., March 1969).

69. Petroleum Resources Under the Ocean Floor, op. cit., pp. 6, 15-19, 88-91.

70. Ibid., pp. 7, 21-25.

71. Ibid., pp. 8-9, 47-53, 92-95.

72. Ibid., p. 10.

73. Ibid., p. 11. According to the NPC the "entire submerged portion of the continent" includes the continental shelf, the continental slope, and at least the landward portion of the continental rise.

74. Ibid., p. 70.

75. Ibid., p. 12.

76. Ibid., p. 56. The NPC report claims that support for its point of view is found in the preparatory work leading to the Continental Shelf Convention, see pp. 57-71. A different point of view is expressed by Professor Carl Auerbach in "Tentative Recommendations of the Commission on Marine Science, Engineering and Resources," Proceedings of the Third Annual Law of the Sea Institute, p. 6. Auerbach claims that there is little evidence that the convention's definition of the continental shelf was intended to include the continental slope.

77. American Petroleum Institute, "Statement of Policy on Jurisdiction Over the Natural Resources of the Ocean Floor," November, 1969.

78. Oil and Gas Journal, November 18, 1968, p. 97. See also the editorial appearing in the May 12, 1969 edition on p. 105.

79. Senate Interior Committee, "Outer Continental Shelf," part II, 91st Congress, 2d session, p. 397. On the oil import quota program and foreign tax credits see the testimony of Thomas F. Field and Professor David Freeman before the Joint Economic Committee, "Oil Prices and Phase II," 92nd Congress, 1st session, pp. 30-41 and 41-52. On the general offshore tax situation see the letter of E. F. Kilpatrick, Director, Division of Tariff Classification Rulings to J. L. Goodier of United Aircraft Research Laboratories in House Report No. 999, 90th Congress, 1st session, p. 111.

80. See, for example, the statement of Cecil Olmstead of Texaco, Senate Commerce Committee, "Special Study on United Nations Suboceanic Lands Policy," 91st Congress, 1st session, p. 100.

81. House Committee on Foreign Affairs, "The United Nations and the Issue of Deep Ocean Resources," 90th Congress, 1st session; Senate Foreign Relations Committee, "Governing the Use of Ocean Space," 90th Congress, 1st session; House Committee on Foreign Affairs, "The Oceans: A Challenging New Frontier," 90th Congress, 2d session; Senate Commerce Committee, "Special Study on United Nations Suboceanic Lands Policy," 91st Congress, 1st session; Senate Interior Committee, Special Subcommittee on the Outer Continental Shelf, "Outer Continental Shelf," 91st Congress, 1st and 2d sessions; and the House Armed Services Committee, Subcommittee on Seapower, "Territorial Sea Boundaries," 91st Congress, 2d session.

82. Senate Foreign Relations Committee, "Activities of Nations in Ocean Space," op. cit., p. 31.

83. Ibid. See also the testimony of Assistant Defense Secretary Warren Nutter on p. 235.

84. Ibid., p. 134.

85. Senate Commerce Committee, "Special Study on United Nations Suboceanic Lands Policy," op. cit., p. 88.

86. Senate Interior Committee, "Outer Continental Shelf," op. cit., pp. 252-253, 263.

87. Ibid., pp. 263-264.

88. Ibid., p. 129. The 1968 Joint Report of the Sections of Natural Resources Law, International and Comparative Law, and the Standing Committee on Peace and Law Through the United Nations of the American Bar Association observed that "the 'exploitability' factor of the Convention is limited by the element of 'adjacency.' The exclusive sovereign rights of the coastal nations to the exploration and exploitation of the natural resources of the seabed and subsoil encompass 'the submarine areas adjacent to the coast but outside the area of the territorial sea.' According to this view [described in the report as the "better view"], therefore, the exclusive sovereign rights of the coastal nations with respect to the seabed minerals now embrace the submerged land mass of the adjacent continent down to its junction with the deep ocean floor, irrespective of depth." The full text of this report appears in ibid., pp. 26-31. Quoted from p. 28.

89. Ibid., pt. II, pp. 299-307.

90. The similarity of points of view has led Senator Pell to claim that "the Interior Department is acting really as the lawyer for the industry in this particular issue." He asserts that "the industry's position is automatically backed up by the Department of the Interior." Additionally, he claims that the ABA attorneys responsible for the report alluded to in footnote 88 are connected with the petroleum industry in their private practices. See Senate Interior Committee, "Outer Continental Shelf," op. cit., part II, pp. 400-401.

91. Senate Commerce Committee, "Special Study on United Nations Suboceanic Lands Policy," op. cit., pp. 120-121.

92. Ibid., p. 124.

93. Ibid., p. 124.

94. Senate Foreign Relations Committee, "Activities of Nations in Ocean Space," op. cit., pp. 221-222.

95. Senate Commerce Committee, "Special Study on United Nations Suboceanic Lands Policy," op. cit., p. 16. That State Department officials gave less credence to the theory of "creeping jurisdiction" was pointed out in interviews with officials of the Legal Adviser's Office.

96. Senate Foreign Relations Committee, "Activities of Nations in Ocean Space," op. cit., p. 218.

97. Ibid., p. 234.

98. Ibid., p. 95.

99. The text appears in Senate Interior Committee, "Outer Continental Shelf," op. cit., part II, pp. 428-429.

100. General Assembly Resolution 2574-D (XXIV), December 15, 1969.

101. Leigh S. Ratiner, "United States Oceans Policy: An Analysis," Journal of Maritime Law and Commerce 2 (1971): 235, 260-261.

102. Senate Interior Committee, "Outer Continental Shelf," op. cit., part II, pp. 430, 433, 455.

103. Ibid., p. 450.

104. Ibid., pp. 433-435, 441.

105. Ibid., pp. 434, 441.

106. Ibid., pp. 450-451.

107. Ibid., pp. 452-453.

108. The full text of this draft convention appears in ibid., part III, pp. 71-96.

109. International Seabed Area Convention, articles 1(2) and 26(1).

110. Ibid., article 27 (2).

111. Ibid.

112. Ibid., article 5(1) and appendix C, article 9.

CHAPTER

7

REACTION TO THE AMERICAN DRAFT TREATY AND THE CURRENT SITUATION

DOMESTIC REACTION

By no means did the May 23, 1970 statement of President Nixon and the subsequent draft treaty reconcile domestic differences on American ocean policy. Indeed, determined and organized opposition met the Administration's position and sought to change American policy. Testifying before the Senate Interior Committee in September of 1970, attorney Northcutt Ely noted fundamental differences between the policy expressed in the proposed seabed treaty and that suggested by the American Bar Association. The disagreement centered around the matter of the "renunciation" by the United States of its existing rights seaward of the 200-meter line; it was the view of the ABA that the United States "assert to the full the rights over adjacent submarine areas now vested in it by the Shelf Convention and by general international law."[1] Ely's arguments against the draft convention reflected nationalistic values and concerns such as the financial loss to the United States Treasury and loss of full national control over adjacent submerged lands. He attacked the creation of what was seen as a new "super-sovereign," observing that under an American regime disputes would be adjudicated in American courts while under the suggested treaty many kinds of disputes would be settled in a new international tribunal.[2] According to Ely "There comes a time when American representatives are trustees for their own people, not for all mankind. It would have been incumbent upon them to say, as perhaps the Russians would say, 'Nyet,' we do not propose to relinquish, renounce, impose a trusteeship on what is ours."[3] In summation he restated the preference of the ABA for national jurisdiction that would allow "Congress to continue to administer the resources of the American continental margin as the heritage of the American people."[4]

The petroleum industry also strongly condemned the draft treaty. American Petroleum Institute spokesman Luke Finlay's criticism revolved around two points. First, the "critical failure" of the draft convention to protect essential United States interests in the mineral resources found in the American continental margins and, second, the "stifling effect" that it would have on the future development of the United States' continental shelf beyond the 200-meter isobath.[5] It was urged that the United States completely set aside its draft treaty and substitute for it a treaty that would "firmly establish the outer limit of coastal nation jurisdiction over seabed resources at the outer edge of the continental margin."[6] The views expressed by the National Petroleum Council echoed those of the API. According to the NPC, the basic approach of the United States was all wrong and had to be changed. The NPC called for a policy reversal asserting that the United States should retain rather than relinquish the "inherent rights" it possessed over the continental margin. The NPC held that "Any treaty which fails to assure effective national jurisdiction over the entire seabed pertaining to the U.S. would be placing in jeopardy a vital national interest of this country."[7]

A spokesman for hard mineral interests, T. S. Ary, Vice-President of Union Carbide and an official of the American Mining Congress, held a different view of the draft treaty since those who were interested in hard mineral exploitation were concerned with deep ocean space rather than with the continental margins. According to Ary, existing legal arrangements did not establish a secure investment climate for the commitment of substantial funds to ocean mining nor did they provide for the security of tenure beyond the limitation of national jurisdiction. It was claimed that the American draft convention represented a tangible step in the direction of more precise international arrangements felt to be needed by the American Mining Congress.[8] Ary however, did criticize the suggested interim policy of granting leases beyond the 200-meter isobath subject to conversion into new licenses under the aegis of the International Seabed Authority. It was claimed that this feature would serve to inhibit the search for minerals in the deep seabed.[9] The draft treaty was further criticized for proposing the establishment of an international seabed authority that was seen as unnecessarily elaborate. It was feared that such a body would create excessive administrative red tape and impose needless financial burdens on those who undertook mineral exploitation. Ary suggested that "simplicity and efficiency without unnecessary expense should be the prime goals of any seabed authority that is established."[10] While this representative of the hard mineral industry found features to criticize in the American draft treaty, his attitude, as opposed to that of representatives of the petroleum industry, was more conciliatory to the basic approach contained in the American draft.

Reaction to the May 23, 1970 announcement of President Nixon was favorable among American oceanographers. In a statement of policy, the Ocean Affairs Board and the International Marine Science Affairs Policy Committee of the National Academy of Sciences endorsed the Nixon policy directive. It urged that the scientific community firmly support the Administration position since it would encourage agreement upon a narrow continental shelf; if widely accepted, it would confine coastal state control over scientific research concerning the shelf to the smallest possible area. In its view, the question of the extent of the continental shelf was too important to be left to resolution by legal interpretation of the Continental Shelf Convention.[11] The scientific community, however, was quite frustrated in its attempts to learn the details of what the American position would be on freedom of research, and it was feared that the United States would compromise on that issue in order to secure agreement on free passage through international straits.[12]

Favorable reaction to the President's policy statement also came from the New York Times[13] and the Commission to Study the Organization of Peace. The CSOP had suggested in 1969 that the United Nations be given jurisdiction over the seabed beyond the limit of 200 meters or 50 miles, whichever was farther from the coast.[14] Nevertheless, the CSOP was pleased that President Nixon called upon states to renounce national claims to seabed areas beyond the 200-meter depth and to regard such resources as the common heritage of mankind. In the way of constructive criticism it suggested that there would be difficulty in creating two types of machinery to authorize resource exploitation beyond 200 meters; the CSOP favored instead the concept of one United Nations authority to regulate all of the seabed past the 200-meter isobath. The elimination of the key concept of the trusteeship area was seen as a measure that would make the American proposal more efficient and more equitable.[15]

THE REPORT OF THE OUTER CONTINENTAL SHELF SUBCOMMITTEE OF THE SENATE INTERIOR COMMITTEE

On December 21, 1970 the report of the Special Subcommittee on the Outer Continental Shelf of the Senate Interior Committee was made public.[16] The views expressed in the report paralleled the position held by the petroleum industry. The subcommittee made major objections to the American draft treaty and indicated that it favored the regime established by the Continental Shelf Convention which, it claimed, allowed exclusive coastal state exploitation to the limit of the continental margin.[17] The draft treaty was interpreted as involving a renunciation of American sovereign rights beyond the 200-meter isobath with the loss of title to the resources of continental margins of the United States. Such a renunciation, it was maintained, in no way guaranteed that the United States would regain from the international community, in the proposed trusteeship zone, the same rights that were renounced. The point was stressed that coastal state sovereign rights to the continental margin were merely reaffirmed by the Continental Shelf Convention and that coastal state control over this area was an inherent right "vested by virtue of the natural extension beneath the sea of our sovereign land territory." Rights in this offshore land were thus not dependent upon the approval of the international community. The draft treaty, in effect, would renounce inherent American rights and request that in part they be returned to the United States.[18]

In an expression of sentiments sounded earlier by petroleum interests, it was noted that access to the oil of the continental margin was of paramount importance to the security of the United States.[19] While to this extent giving consideration to security requirements, the subcommittee displayed little concern with "creeping jurisdiction." Its report observed that the overwhelming majority of coastal states party to the Continental Shelf Convention had limited their jurisdictional claims to the terms of that treaty.[20] In stating this the report failed to note that only 49 states were then party to the Continental Shelf Convention. The major problems had been caused by the larger number of states, such as those of Latin America, which were not then and are not now parties to this convention. It was claimed that the best way to insure that the provisions of the Continental Shelf Convention were obeyed was to seek to enforce them, but insofar as treaty obligations were involved, they required adherence only by those states party to the treaty. According to the subcommittee report the freedom of the seas doctrine was well recognized in customary international law; thus, by implication, it could be held that states not party to the Geneva Conventions on the Law of the Sea were bound by customary international law to respect certain uses of the sea by other states.[21] But even accepting this claim, customary principles of international law are subject to change. Many of the new states, for example, wish to alter important features of customary law since such law was created before the new states had any input into the establishment of these norms. There is strong suspicion that much customary international law benefits the older, more developed states, at the expense of the newer, less developed countries.[22]

Realistically, what could be done to counter the erosion of a customary doctrine of international law such as that of freedom of the seas? The threat or use of force by a large state such as the United States could very well be counterproductive. The alternative is to seek international agreement in treaty form; the achievement of such agreement requires, however, "give and take" rather than unilateral action. In the long run, though, valued rights are made more secure.

Not content with indicating its opposition to the American draft convention, the Interior subcommittee warned that according to its interpretation of the United States Constitution any modification of the property rights of the United States to the mineral resources of the continental shelf would require an act of Congress. According to Senator Metcalf, article IV, section 3 of the Constitution, which states in part that "The Congress shall have Power to dispose of and make all needful Rules and Regulations respecting the Territory or other Property belonging to the United States . . . ," suggests that ratification of the treaty by the United States government would have to be followed by legislation to carry out American obligations under that treaty.[23] The Senate Interior Committee was clearly serving notice that it intended to exercise an important function in shaping American seabed policy. If, in fact, legislation were needed after treaty ratification, it would be channeled through the Senate Interior Committee, whose membership could vitally affect the character of such enabling legislation. In part the emphasis on the significant role of the Interior Committee was motivated by what members of that committee saw as the "muscling in" of the Senate Foreign Relations Committee into an area of Interior Committee competence: offshore lands.[24] A treaty, of course, would pass through the Foreign Relations Committee and allow the impact of that committee to be felt; the Interior Committee wanted to insure a major position of influence for itself. The realization that enabling legislation would be necessary would have significant impact upon the international negotiating position of the United States.

The opposition of the Interior subcommittee did have some immediate impact upon the American draft treaty. The State Department was pressured into labeling the United States Draft Convention on the International Seabed Area a "working paper."[25] Added to the draft convention's face was the following statement:

> The attached draft of a United Nations Convention on the International Seabed Area is submitted by the United States Government as a working paper for discussion purposes.
>
> The draft Convention and its Appendices raise a number of questions with respect to which further detailed study is clearly necessary and do not necessarily represent the definitive views of the United States Government. The Appendices in particular are included solely by way of example.

During hearings in September of 1970, Senator Metcalf emphasized the importance of the above "disclaimer" and insisted that the draft treaty did not have the status of an approved United States government position.[26] It is interesting to note that Assistant Secretary of the Interior Hollis Dole shared this opinion. In a letter dated August 5, 1970, two days after the presentation of the American draft treaty to the UN Seabed Committee, and addressed to the National Petroleum Council, Dole noted that the draft was circulated as a "working paper" and that it "does not necessarily represent the definitive views of the United States Government." Dole requested that the NPC reconvene its group of experts to study the draft convention and to prepare an article by article analysis of the draft treaty as well as alternative wording. Obviously, the Interior Department had not given up hope that significant changes in American policy could still be brought about.[27]

INTERNATIONAL REACTION

On August 3, 1970 the United States draft treaty was submitted to the UN Seabed Committee for its consideration. According to the United States representative, the draft had been drawn up after "painstaking examination of relevant national and international interests." It was asserted that the convention involved an entirely new approach to the law of the sea providing a framework for the exploitation of seabed minerals and, very importantly, "for the first time in history the equitable division of the benefits of that exploitation, regardless of the advantages of technology or geography enjoyed by any State."[28]

Despite the belief on the part of the American delegation that it had presented a draft treaty worthy of extended consideration, containing many features which would be attractive to other states, the American proposal failed to capture the imagination of the Seabed Committee. Some states such as the Soviet Union and Chile felt that it was too soon to consider a specific regime for ocean areas and that there was first a need to develop general principles. The representative of the Soviet Union suggested a more "realistic" approach rather than that of the United States, which reflected "unjustified optimism." More information was said to be needed before any definite rules and machinery could be established.[29] Some states favored a simpler division of the seabed between national areas and those areas under an international regime.[30] The concept of the trusteeship zone came under attack. According to the representative of India:

> that area would constitute between 25 and 30 percent of the sea-bed area, where most of the known petroleum and mineral resources might be available. If each State became a trustee for the continental margins, there would be greater scope for the development of a monopoly or trusts by a few technically advanced countries which alone had the necessary skills, equipment and resources. In effect, therefore, some foreign companies rather than the coastal State concerned would become trustees.[31]

The representative of Kenya stated that the trusteeship area beyond the 200-meter limit, including the continental slope, was "unacceptable." He noted that it was precisely this area which was rich in natural resources; the international authority, though, was given exclusive jurisdiction only over the deep seabed.[32] The Peruvian delegation suggested that the United States plan took account primarily of United States interests and claimed that the less developed states had little to gain from adoption of it. It was said that the net profits of the international authority would be:

> the amounts grudgingly given in the form of licenses, taxes and other charges; that a part of that revenue would be allocated for the administrative expenses of the organization; that another part would be used to promote the efficient exploitation, investigation and protection of the marine environment. Only when all that had been deducted would the balance be used to encourage the economic progress of the developing countries.[33]

Finally, the point was made by the representative of Iceland that there was a need to deal with the matter of fisheries as well as that of mineral resources.[34]

Aside from the specific criticisms of the American draft treaty, an analysis of the observations of foreign diplomats on the proposal indicates that it was regarded as too refined and too definite a plan for the period of negotiations at which it was introduced. It seemed to reflect the belief that wide consensus already existed or was close to achievement. Its carefully balanced articles, protecting American interests yet emphasizing benefits for the international community served as an irritant to some. Why then did the State Department put forward such a detailed draft convention at this point in time? According to State Department officials who were interviewed, there was strong domestic political pressure to submit a definite program to the UN Seabed Committee as soon as possible after President

Nixon's May 23, 1970 statement. There was fear in some quarters, from the Defense Department in particular, that the presidential decision might be changed, since powerful political forces were opposed to the direction of presidential leadership and were actively seeking a policy reversal. With this consideration in mind it was decided to "strike while the iron was hot" and act upon the presidential decision in the hope that this would make it more difficult for the President to later reverse himself. In an important sense then, the presentation of the August 3, 1970 draft treaty was aimed at locking the United States into a particular policy. While this may have served certain domestic political purposes, the timing of the American presentation internationally was not optimum since sufficient international consensus had not yet been created.

The United States, in an attempt to counter the above mentioned specific criticism, indicated that various "misconceptions" had led some delegations to oppose the trusteeship zone. It was suggested that the principle of such an area was confused with the specific limits proposed by the United States. The point was made that American thinking as to the extent of the zone was flexible and the particular mixture of coastal and international rights and duties was clearly negotiable.[35] With respect to the matters of fisheries and territorial waters, the United States on August 3, 1971 submitted Draft Articles on the Breadth of the Territorial Sea, Straits, and Fisheries.[36] The introduction of these new draft articles represented a shift in United States thinking on ocean problems. According to State Department Legal Adviser Stevenson, the United States government had originally believed that consecutive treatment of the different issues of the law of the sea would be the best approach;[37] the expression of views by other states, however, had led the United States to recognize the need to resolve important ocean issues simultaneously. All this, though, was to no avail as an attempt to save the viability of the American draft convention on the International Seabed Area.

THE EMERGENCE OF THE ECONOMIC ZONE CONCEPT

In 1972 an alternative approach to the regime for ocean space began to emerge; its key recommendation called for the creation of an offshore economic zone which would be controlled exclusively by the coastal state. The interest of the UN Seabed Committee was aroused by developments that occurred at the Santo Domingo Conference of Caribbean Countries on the Problems of the Sea and at the African States Regional Seminar on the Law of the Sea. Both these meetings took place in June 1972 and both produced proposals embodying an offshore economic zone.

The Declaration of Santo Domingo introduced the concept of the "patrimonial sea." In the area so designated the coastal state was to have sovereign rights over the renewable and nonrenewable natural resources found in the water, seabed, or subsoil. It was also to possess the right to regulate scientific research and the power to take measures to prevent marine pollution in this area. While the extent of this zone was to be determined by international agreement, it was felt that it should not exceed a breadth of 200 miles from the coast. Very significantly, in the patrimonial sea, ships and aircraft of all states were entitled to freedom of navigation and overflight with no restrictions "other than those resulting from the exercise by the Coastal State of its rights within the area." In its statement on the continental shelf, the declaration accepted the Continental Shelf Convention's definition of the extent of the shelf but noted that the portion of the shelf that was located within the patrimonial sea was to be subject to the regime provided for that area; that part of the shelf past the patrimonial sea would remain subject to existing international law. The seabed beyond the patrimonial sea and the continental shelf was to be treated as the common heritage of mankind.[38]

The African States Regional Seminar on the Law of the Sea reached similar conclusions recommending the creation of an economic zone. In this zone the coastal state would have exclusive jurisdiction for the purposes of owning, controlling, and regulating the national exploitation of the living and nonliving resources of the sea and seabed. Within it the coastal state was also to have the power to prevent and control pollution but the zone was to be created without prejudicing freedom of navigation, overflight, or the laying of submarine cables or pipelines. No mention was made of freedom of scientific research and no definite extent for the economic zone was suggested, although it was recommended that such an area encompass at least the continental shelf. The natural resources beyond this zone were to be managed by an international authority which would have special concern for the promotion of "the progress of mankind in the developing countries so as to correct the grave imbalance between the nations."[39]

Unlike the American draft treaty, these proposals did have great and immediate impact upon and appeal to a number of states, which believed that the economic zone concept provided a proper basis for the drafting of a new regime for ocean space. It was felt that these proposals would allow for the utilization of marine resources, provide a share of the ocean's wealth for the less developed states, and minimize interference with the legitimate interests of other states while protecting the interests of the coastal states in nearby offshore areas.[40]

Accordingly, in 1972 Kenya introduced for consideration by the UN Seabed Committee a proposal for the creation of an economic zone.[41] The Kenyan draft articles would recognize the right of all states to establish an economic zone beyond the territorial sea. Within this zone the coastal state would have sovereign rights over natural resources and would have exclusive jurisdiction for the purposes of control, regulation, and exploitation of both living and nonliving resources and for the purpose of prevention and control of pollution. The establishment of the economic zone was not to affect freedom of navigation, overflight, and the laying of submarine cables and pipelines "as recognized by international law." The coastal state would be authorized, however, to regulate scientific research. The envisaged extent of the zone was not to be in excess of 200 nautical miles from the baseline of the territorial sea.

In an important statement on August 10, 1972, American spokesman John Stevenson told the UN Seabed Committee that the United States would be willing to accept coastal state jurisdiction over an economic zone beyond the territorial sea as a part of an overall law of the sea settlement. American acceptance of an economic zone, an act that formally signaled the death of the 1970 United States draft treaty, which had long been dying of neglect, was subject to five significant conditions. First, standards would have to be created and accepted so as to prevent unreasonable interference with uses of ocean space not involving resource exploitation. Second, international treaty standards to protect the ocean from pollution would have to be accepted. Third, the new treaty on the law of the sea would have to insure integrity of investment. Fourth, the United States would insist that there be some sharing of revenues from an economic zone for international community purposes, particularly for the benefit of developing states. Fifth and finally, there would have to be a system of compulsory dispute settlement to protect noncoastal and international interests.[42]

The American draft treaty of August 1970 had failed to impress the coastal states of the world because it did not provide these states with the degree of control in offshore areas that they desired. Since the basic concern of the United States in law of the sea negotiations was to achieve implementation of the five principles mentioned above and since it was apparent that the concept of the economic zone was favored by a large number of states and that the United States could realize its goals in conjunction with the creation of an economic zone, the United States abandoned support for the less popular trusteeship zone.[43] This change signified that nationalism as a value would be reflected to a greater degree in the developing regime for ocean space than if the American plan of 1970 had been accepted.

In 1973 proposals for an economic zone were incorporated into draft treaties and submitted to the UN Seabed Committee for its consideration. Colombia, Mexico, and Venezuela introduced a draft treaty that would create a patrimonial sea in accordance with the recommendations of the Declaration of Santo Domingo.[44] In May of 1973 the Council of Ministers of the Organization of African Unity met to harmonize the position of African states on law of the sea matters and, at their conference in Addis Ababa, recognized the right of each coastal state to establish an exclusive economic zone beyond its territorial waters to an outer limit of 200 miles as measured from the baselines for its territorial seas. In such zones the coastal state would exercise "permanent sovereignty" over living and mineral resources and would administer the zone without "undue interference" with freedom of navigation, overflight, and the laying of cables and pipelines. Scientific research within the economic zone, however, was to be carried out only with the consent of the coastal state. Ocean areas beyond national jurisdiction, according to the OAU, were to be dealt with as a common heritage of mankind and managed by international machinery. The African states favored a strong international governing body, which could itself explore and exploit the area and could minimize any adverse economic effects of price fluctuations of raw materials caused by activities carried out in the sea. Presumably then, this body would possess power to control levels of deep sea exploitation. Additionally, the governing body was to be authorized to regulate and to undertake scientific research and to protect deep ocean areas against marine pollution. The international machinery that would govern this area was envisaged to include an assembly of all member states and a council of limited membership, reflecting the principle of equitable distribution and exercising in "a democratic manner" most of the functions of the machinery. There was also to be a secretariat and a tribunal for the settlement of disputes.[45] The conclusions of the OAU were then embodied in a 1973 draft convention cosponsored by 14 African states and submitted to the UN Seabed Committee.[46]

On July 16, 1973 the United States introduced its draft articles on a coastal seabed economic area.[47] Unlike the aforementioned proposals for a coastal economic zone, the American draft would grant exploration and exploitation rights solely to seabed resources; fisheries were not dealt with in this plan. The extent of the coastal seabed area was left open for future negotiations but it was noted by the American delegation that the preponderant view held by representatives of other states was that the outer boundary of an economic zone should be at a distance of 200 miles from the coast or should be marked by the outer boundary of the continental margin if it extended beyond 200 miles.[48] The American proposal also incorporated stipulations that would give effect to the five conditions insisted upon by Ambassador Stevenson in his August 10, 1972 address to the Seabed Committee.

While the United States was moving toward a position on the law of the sea that was more in line with the views of a majority of states, significant differences still existed between American preferences and those of the states favoring an offshore economic zone. Disagreement was evident in at least four areas: fisheries, pollution control, the regime for deep ocean space, and scientific research.

With respect to fisheries, the African proposal for an economic zone and that of Colombia, Mexico, and Venezuela for a patrimonial sea envisaged exclusive coastal state exploitation rights to fishery resources in these zones. American preferences, however, reflect a rejection of a zonal approach to fishery regulation and opt, instead, for a species approach. As seen in the August 1972 United States draft articles on fisheries, three categories of species are established, each to be regulated in accordance with its biological characteristics and in such a manner so as to assure its conservation, maximum utilization, and equitable allocation.[49] Coastal stocks of fish are to be managed by the coastal state, which is to have preferential rights to these fish, even in areas beyond the territorial sea all the way to the limits of their migratory range. Anadromous species, which like salmon spawn in fresh water and then migrate far out to sea, are also to be regulated by the state in which they spawn to the complete extent of their migratory range. The state involved would again have preferential rights to these species throughout its range. To insure maximum utilization and equitable allocation of coastal and anadromous resources, the coastal state is entitled to reserve for its own fishermen the portion of the catch that they are capable of taking. If there remain nonutilized fishery resources the coastal state shall then provide access to other states, with fishermen from other states that have traditionally fished in the area being given preference. The third category of fishery species was that which was highly migratory; control of such fish, in line with the American proposal, would be delegated to appropriate international fishery organizations rather than to individual states. Tuna was the primary example of such a fish.

On the matter of pollution control, both the African proposal and that of Colombia, Mexico, and Venezuela would grant the coastal state regulatory power to protect against marine pollution in the economic zone. The lack of any elaboration on this right of the coastal state proved discomfiting to the United States, which wanted coastal state powers made clear and limited by international agreement. While favoring action on ocean pollution, the United States did not want ocean pollution control to afford an excuse for interference with other uses of the seas. The American delegation noted that the problem of pollution emanating from ships was fundamentally different from that posed by land based sources or seabed resource activities and, consequently, it required a different solution. Ambassador Stevenson observed that vessel pollution necessitated a careful balancing of interests, rights, and obligations among maritime and coastal states. He stressed that control of such pollution must be based on international agreement. It was suggested that providing great discretion and power to the coastal state would not lead to a proper balance and indeed, could lead to serious conflict.[50]

In July of 1973, the United States submitted to the Seabed Committee draft articles on the protection of the marine environment.[51] The American plan would give to the International Seabed Resource Authority power to establish standards for operations within the coastal seabed zone and in the area seaward of it. The coastal state, however, would be entitled to set up stricter standards for activities in its seabed economic area. Development of rules on vessel pollution would be the responsibility of the Intergovernmental Maritime Consultative Organization; IMCO, it was asserted, should be given this task because of its technical competence. In the coastal seabed zone the coastal state would have complete authority to enforce national and international standards in relation to seabed activities, subject only to international inspection to insure compliance with internationally set standards.

In the American view, responsibility for coping with pollution from ships would be shared by flag states, port states, and coastal states. The flag-state would be obliged to enforce international rules against its own ships. Port states could also enforce these rules against ships using their ports regardless of where the violation occurred. Lastly, the coastal state would be given the right to take "reasonable emergency enforcement measures" as were necessary in the area beyond the territorial sea to "prevent, mitigate, or eliminate imminent danger of major harmful damage to its coast or related interests from pollution arising from a particular occurrence reasonably believed to be related to a violation of the applicable international standards."[52]

To minimize the possibility of costly delays resulting from prolonged detention of ships, the United States draft provided that vessels should not be held longer than necessary for investigative purposes. In cases where investigations indicated that a violation had occurred, ships were to be released promptly after bonding except in the case where such release "would present an unreasonable threat of harm to the marine environment or where other action is required or authorized by the applicable international standards."[53] The American draft further included a provision for compulsory dispute settlement.

It was suggested by American spokesmen before the Seabed Committee that the proposed system would be effective in protecting the ocean environment while at the same time protecting shippers against unreasonable delay. Article 21 of the proposed draft specifically noted that "A State shall be liable for damage resulting from investigative, enforcement or intervention measures exceeding those reasonably necessary in the light of available information."

It is interesting to note that while in the past it had been the United States Navy that was deeply concerned with navigation rights as a consequence of the development of special purpose zones, with interest growing in relation to pollution control the petroleum industry began to show a new interest in freedom of navigation. Whereas in the past the National Petroleum Council felt that existing international law provided sufficient protection for navigation, doubts began to appear as seen in its May 1973 report and more recently its report of July 1974.[54] While it was admitted that the coastal state had important interests in protecting the marine environment, it was urged that these interests be clearly defined and that internationally agreed upon rules be established, which would protect the environment without hampering navigation. It was emphasized that the economical transportation of petroleum required that oil tankers pass through coastal waters and through principal international straits; interference with tanker traffic in the name of pollution control would have important negative impact upon the whole world. In its latest report the NPC seems to have moved away from its total rejection of the concept of "creeping jurisdiction," for it now noted that:

> Another way of attempting unilaterally to extend the jurisdiction of a Coastal State over navigation, referred to by the Canadian Delegate to the U.N. Seabed Committee as 'functional jurisdiction,' is the attempted exercise of Coastal State authority over defined activities taking place beyond the territorial sea. Such exercise of Coastal State jurisdiction has been frequently undertaken with respect to national defense and security, the protection of fishing interests, and pollution control. The exercise of such jurisdiction has already led to serious impairment of navigation.[55]

Consequently, the NPC has asserted that a principal objective of the United States in law of the sea negotiations should be to obtain international agreement insuring for merchant vessels of all states the right of unimpeded navigation subject to internationally defined rules and dispute settlement procedures. The 1973 United States proposal on pollution control attempts to limit the chance of interference with navigation rights.

With respect to the regime for areas beyond the economic zone key questions that will require answers include: what kind of powers will an international authority possess? What kinds of institutions and organs are needed? How will influence in such bodies be distributed? What payments should be made to an international authority? One point upon which there does seem to exist general consensus is that at least the resources of the seabed and subsoil beyond the limits of national jurisdiction are the common heritage of mankind. Beyond this there is sharp disagreement on a number of matters as reflected in the report of Subcommittee I of the UN Seabed Committee.[56] According to the 1973 report, among the differences that exist is the question of how to define "resources" in the area to be regulated or controlled by the international regime; is the term to apply only to nonliving resources?[57] Of great importance is the matter of who will be entitled to exploit resources in the international area. Some favor exploitation by states and by private enterprise, while others support a policy whereby all exploration and exploitation of resources and scientific research would be conducted by the international seabed authority itself.[58] On this point, State Department officials have claimed that as preparatory work has progressed it has been increasingly accepted that both public and private entities would be allowed to undertake operations; it is said that there is growing recognition that the seabed authority would not be able to attempt exploitation itself. As a result this issue has been narrowed down to whether the authority will issue licenses on a nondiscriminatory basis or negotiate contracts and joint ventures with discretion and the authority to pick and choose among investors. The United States government strongly favors a system of nondiscriminatory access to exploitation rights in accordance with conditions explicitly stated within the text of the treaty being drafted.[59]

Of very great concern to the United States is the proposal of a number of states that the international authority be empowered to regulate the production, marketing, and distribution of raw materials derived from the seabed whenever it believes that such production or marketing may have adverse effect on developing states that produce and export the same raw materials.[60] This suggestion has been vigorously opposed by the United States since it is seen as being detrimental to the rights of access and to stability of investment. Such power in the hands of the authority, it is claimed, would result in higher prices to the consumer.[61] Additionally, the point can be made that insofar as the United States views the deep ocean bottom as a secure source of hard minerals the hope is dimmed by this proposal for, once more, the availability of raw materials becomes subject to stoppage by a political decision. The Arab oil embargo of 1973-1974 would tend to underscore this concern.

In the area of scientific research within the economic zone, the draft treaties of Colombia, Mexico, and Venezuela and that of the African states would authorize the coastal states to regulate such research.[62] On July 19, 1973 the United States introduced a set of draft articles on marine scientific research[63] which, according to American spokesman Donald McKernan, provided a proper balance between the legitimate concerns of the coastal state and the need for scientists to carry out their research with a minimum of restraints.[64] States were to promote and facilitate scientific research but, to win support among less developed coastal states, the American draft articles created certain obligations for researchers working within the resource jurisdiction areas. The researching state would have to provide advance notification, certification of the project as one which involves only scientific research, opportunity for coastal state participation, a share of samples for the coastal state, and publication of findings. Furthermore, the state undertaking the research would have to comply with applicable international environmental standards and would be obligated to assess the implications of its findings from the point of view of the interests of the coastal state. In case of any differences as to the interpretation or application of the draft articles, either party to a dispute could make use of compulsory settlement procedures which were to be provided.

The approach of the United States to the problem of scientific research may be seen as a compromise between suggestions that would give greater weight to the interests of coastal states and those that would grant greater freedoms and fewer responsibilities to scientific researchers. The 1973 proposal of a group of East European states, for example, called for recognition of the right of scientific research for all states in what was termed the "world ocean." This area was defined as including all of ocean space, the seabed and subsoil thereof, with the exception of internal and territorial waters, and the seabed and subsoil of the continental shelf.[65] While scientists would derive certain benefits from this plan there was little incentive for coastal state acceptance since only vague and general provisions were contained on the subject of benefits to the coastal state. On the other hand, states such as Canada[66] and, as mentioned earlier, some of the Latin American and African states insist that marine scientific research in areas within the jurisdiction of a coastal state shall be conducted only with the consent of that state.

In light of such differences and the growing pressure for coastal state control in offshore areas, an official of the National Academy of Sciences who was interviewed expressed the belief that the adoption of the United States proposal on scientific research was the best outcome that could be realistically expected from law of the sea negotiations. The point was emphasized that the small portion of the scientific community with an interest in law of the sea developments had little political strength and lacked the economic power of other interests concerned with shaping the future regime of ocean space.

While important differences thus exist in connection with the economic zone concept, it is clear to State Department officials that some type of economic zone will be a part of the future legal framework of ocean space. "Nationalists" like Northcutt Ely[67] and spokesmen for the oil industry[68] can take comfort in the emerging zone since it would insure coastal state control over the continental margin, while some who are more internationally minded, such as Senator Pell, indicate continued preference for a plan involving a trusteeship zone.[69]

INTERIM POLICY

The fact that there are a large number of interrelated and complex issues to be treated in a new round of law of the sea negotiations has led to the widespread belief that a comprehensive treaty on ocean space will not be forthcoming for some years. A number of observers have suggested that many states which are exporters of hard minerals such as manganese, copper, nickel and cobalt,(minerals found in ocean bottom nodules), and exporters of petroleum are purposely prolonging discussions so as to delay as long as possible the development of new sources of these commodities, which would compete with their own. It is maintained that this attitude was reflected in the 1969 General Assembly Moratorium Resolution that sought to prohibit the exploitation of mineral resources beyond the area of coastal state jurisdiction. Consequently, it is suggested that unless there is very strong pressure brought by states like the United States for an international treaty none will, in fact, result.[70]

In 1972 and 1973 several bills were introduced in Congress to allow for the development of hard mineral resources found in the deep seabed.[71] The effect of these bills would be to authorize the Secretary of the Interior to issue licenses for deep seabed mineral resources that would be exclusive as against all persons subject to the jurisdiction of the United States or of any reciprocating state.[72] A number of reasons were presented in support of passage of such legislation.

The American Mining Congress, which had drafted the proposed legislation, was of the belief that it was unlikely that a treaty on the law of the sea would be ratified during the 1970s. Technology had progressed at a quicker pace than had the law with the resulting effect that industry, which possessed the technological capability to tap the minerals of the deep seabed, was frustrated by the lack of a protective legal framework. In hearings on a Deep Seabed Hard Mineral Resources Act, spokesmen for Tenneco and its subsidiary, Deepsea Ventures, noted that their companies had invested $20 million to develop the ability to recover and convert the minerals found in manganese nodules into metal and were now ready to undertake the next phase of their work, namely defining a mine site in detail and tailoring engineering designs to the specific physical conditions of the mine site. Deepsea Ventures had already located several potential sites and was ready to register a claim to any of these areas. If this could be done in accordance with legislation of the type being considered, that company was prepared to spend $16 million over the next two years in examining the ore body and in working out the necessary engineering design to conform to the needs of the particular mining area. At the end of this period another expenditure of $150 million was planned for capital construction. It was believed that within three years of this investment metal could be delivered to market. According to the estimates of Deepsea Ventures, annual production of metals would be 500 million pounds of manganese, 25 million pounds of nickel, 20 million pounds of copper, 4.5 million pounds of cobalt, and 5.5 million pounds of other metals including molybdenum, vanadium, and zinc.[73]

In seeking political support for the passage of the Deep Seabed Hard Minerals bill, the hard mineral industry and its supporters noted the important advantages that could accrue to the United States through enactment of this legislation. It was asserted that the development of alternative sources of minerals from the ocean floor would lessen America's dependence upon developing states "which are generally unresponsive to U.S. foreign policy or which are increasingly combining among themselves to manipulate prices upward or to control production and distribution to further their economic and political goals."[74] C. H. Burgess of Kennecott Copper and vice-president of the American Mining Congress warned of the impact of the creation of OPEC-like bodies such as CIPEC (the Council of Copper Exporting Countries), on developed states.[75] According to figures supplied by Tenneco, the United States imports 19 percent of its copper, 84 percent of its nickel, 92 percent of its cobalt, and 98 percent of its manganese.[76] Ocean bottom exploitation, it was said, would free the United States of such dependency. Illustrating the great potential of ocean space in this regard is the judgment of F. L. LaQue, former vice-president of the International Nickel Company, who claimed that 1 percent of the ocean bottom could satisfy the 1967 level of world needs for manganese, nickel, copper, and cobalt for about 50 years.[77]

Not only would the United States gain more reliable sources of needed minerals, thus increasing the freedom of choice in deciding

American foreign policy, but seabed exploitation would strengthen the economy of the United States in several significant ways. It is suggested that the existence of alternative mineral resources would help stabilize their prices at reasonable levels. The United States, with seabed minerals available to it, would be able to negotiate with mineral exporting states from a position of strength.[78] Additionally, the point is forcefully made that seabed exploitation has great potential for improving the American balance of trade by reducing the need for the purchase of mineral resources abroad.[79]

In July of 1973, Dr. Robert White, administrator of the National Oceanic and Atmospheric Administration of the Department of Commerce, asserted that in determining American policy in ocean space the "ocean balance-of-payments value" should be considered. Noting that in calendar year 1972 the United States balance-of-payments deficit was $10.3 billion and the American trade deficit stood at $6.9 billion, he claimed that Commerce Department figures indicated that the 1972 "ocean balance-of-payments" deficit amounted to more than $8 billion. What Dr. White was saying was that the United States was importing ocean products and services and also certain raw materials, which could instead be gained by ocean exploitation, at a rate that contributed more than $8 billion to the total $10.3 billion American balance of payments deficit. Clearly, meaningful ocean exploitation by the United States, he suggested, would significantly affect these figures.[80] State Department officials who were queried about the views of Dr. White agreed that to the extent that seabed minerals were extracted and processed by American industry there would be a definite favorable impact upon the United States balance of payments, but they seemed to downgrade Dr. White's analysis, noting that the concept of an ocean balance of payments had "deficiencies."[81]

It is further claimed that seabed exploitation will directly and indirectly increase employment and thus the economic well-being of the United States. The Shipbuilders' Council of America, for example, supported passage of the Deep Seabed Hard Mineral Resources bill because of the advantageous effect that it would have on American shipyards.[82] Finally and importantly, in the opinion of observers like John Laylin, favorable consideration of this bill would serve as a strong inducement to other states to negotiate in earnest for the establishment of a new regime in ocean space.[83]

Significantly, the American Mining Congress and other supporters of this bill portrayed the bill as an "interim measure," the need for which was suggested in President Nixon's May 23, 1970 declaration on ocean policy. It was noted that enactment of this legislation was for the purpose of filling a void later to be filled by a regime for ocean space created by treaty. The Deep Seabed Hard Mineral Resources bill contained a provision according to which licenses issued under the authority of this act were to be made subject to any international regime for the deep seabed agreed to by the United States subject to two provisions. These were that the international regime would have to recognize fully and protect the exclusive rights of each licensee to develop the licensed block for the term of the license, and that the United States would have to reimburse fully the licensee for any loss of investment or increased costs incurred by the licensee within 40 years after issuance of the license due to requirements or limitations imposed by the international regime which were more burdensome than those contained in the Deep Seabed Hard Minerals bill.[84]

While some called for the passage of this bill as a necessary interim measure that would have the effect of speeding up international negotiations for a permanent governing framework for the deep seabed, others strongly attacked it as an attempt to legitimize the unilateral appropriation by the United States of a portion of what was considered by the international community to be "the common heritage of mankind."[85] Passage of such legislation, it was asserted, would encourage further appropriation of seabed areas by coastal states. According to Professor Wolfgang Friedmann "the bill would complete the process of the gradual partition of the sea in that area which has hitherto under most extensive interpretation been regarded as free to all nations."[86] As to the effect of its passage upon international negotiations, Professor John Logue has stated:

> We are saying we will take any regime that the U.N. comes up with, provided first, that it ratifies every license we have given; and, second, that if the U.N. does not ratify those U.S.-granted licenses, the Congress of the United States must make good the losses of any licensee.
>
> My guess is that the U.N. would not buy the first proposition, and the U.S. Senate would not buy the second. The U.S. Senate would not agreed [sic] to bail out all the concerns whose licenses might be jeopardized by or . . . be regarded as completely invalid by the international regime.[87]

What is being charged here, then, is that such proposals for "interim" legislation, because they would prove to be internationally unacceptable, would, in fact, become permanent. That other states were concerned with the possible passage of this legislation is seen in their criticism of the United States on the grounds that consideration of such legislation was prejudicial to law of the sea negotiations.[88]

The position of the State Department on Deep Seabed Hard Mineral Resources bills has been cautious. In March of 1973 Acting Legal Adviser Charles Brower noted that the legislation "is similar to unilateral claims that we oppose and that are contrary to our security, navigation and resource interests, and moreover preempts the Law of the Sea Conference on this issue." The State Department had urged other states to refrain from unilateral claims in favor of participation in a multilateral negotiation process and it was strongly maintained that the United States should do the same "so long as there are reasonable prospects for a timely and successful conference." Brower indicated that the bill being considered did not "appear" to be satisfactory but nevertheless agreed that prudence dictated that the United States begin to formulate a legislative approach to operations in the deep seabed in case negotiations failed to come to a "timely and successful" conclusion. According to Brower, "timely" meant that a convention, which would provide for the provisional coming into effect of the international regime,[89] would be open for signature in 1974 or 1975 at the latest.[90] By taking such a position the State Department is clearly exerting pressure on other states to agree to a new law of the sea convention; the United States is serving notice that American exploitation of the deep seabed will occur either in accordance with a new treaty, as the State Department prefers and in the negotiating of which other states would have considerable influence, or through unilateral action by the United States. The United States will not stand for what it regards as unwarranted delay.

Growing American impatience and pressure for immediate action is demonstrated not only in relation to seabed exploitation but also with respect to fisheries. Several bills were reintroduced in Congress in 1973 calling for the establishment of an exclusive American fishery zone extending 200 miles out to sea.[91] This action reflected the pessimism of northeastern United States fishermen about the possibility of achieving international agreement that would safeguard their livelihood.[92] More significant than the introduction of these bills is the passage of the 1973 Offshore Shrimp Fisheries Act.[93] This act, which essentially constitutes enabling legislation for the carrying out of obligations under the 1972 Brazil-United States shrimp treaty,[94] is important because of an amendment to it which makes the lobster, Homarus Americanus, a creature of the continental shelf. In line with the Continental Shelf Convention, this gives the United States exclusive management over this resource. The amendment was specifically designed to protect lobster found beyond the 12-mile exclusive fishing zone of the United States. Approximately 25 percent of the lobster catch occurs beyond this limit. Both the State Department and the Department of Commerce expressed doubts about the wisdom of adopting this amendment in view of the fact that international negotiations were under way in an attempt to solve the problem caused by foreign fishing off American coasts.[95] The amendment was approved after the Commerce Department, which now has the power to define continental shelf species, refused to so classify lobster because it could swim and thus, biologically, did not qualify as a shelf resource. Further, both State and Commerce maintained that by declaring lobster a resource of the shelf other states might be encouraged to take similar action with respect to species such as shrimp and spiney lobster—species that American fishermen take off foreign shores.[96] These arguments went unheeded and the amendment became law. Though the State Department was cautioning other states to avoid taking unilateral action in connection with ocean areas, this was precisely what the United States government had just done. The significance of this action remains to be seen.

In summation, the current period is of major importance in deciding the future regime of ocean space. Either ocean space will be developed on the basis of an internationally agreed upon framework or in accordance with a series of unilateral actions that will give further accent to national control. It is very clear that strong pressures, both nationally and internationally, have forced a change away from the August 1970 American draft treaty which embodied the trusteeship concept and, as compared to the evolving concept of the economic zone, reflected greater willingness to recognize a more meaningful role in ocean space for the entity commonly referred to as the international community.

NOTES

1. Senate Interior Committee, "Outer Continental Shelf," op. cit., part III, 91st Congress, 2d session, pp. 4-7.
2. Ibid., p. 12.
3. Ibid., p. 19.
4. Ibid., p. 13.
5. Ibid., p. 44.
6. Ibid., pp. 48-49.
7. National Petroleum Council, <u>Petroleum Resources Under the Ocean Floor: Supplemental Report of the National Petroleum Council</u> (Washington, D.C., 1971), pp. 15-16, 31.
8. Senate Interior Committee, "Outer Continental Shelf," op. cit., part III, pp. 37, 39.
9. Ibid., p. 39. This point was strongly underscored by a spokesman for the American Petroleum Institute, see p. 150.
10. Ibid, p. 40.

11. National Academy of Sciences, A Statement of the Ocean Affairs Board and the International Marine Science Affairs Policy Committee, "Continental Shelf Limit and Freedom of Scientific Research in the Region Beyond," n.d.

12. On the continuing frustrations of the scientific community see, New York Times, December 26, 1971, p. 48. See especially the comments of William Burke and Warren Wooster.

13. New York Times, August 7, 1970, p. 32. The editorial stated in part that "The proposed international agency would not, as has been charged, deny to Americans their fair share of seabed resources. Rather it would provide a more stable regime under which American entrepreneurs, with their superior technology, could operate for reasonable private gain and for the benefit of mankind. President Nixon deserves strong public support for this promising initiative. It puts the United States once again in the forefront of efforts to promote a more orderly and just international community."

14. Commission to Study the Organization of Peace, "The United Nations and the Bed of the Sea," Nineteenth Report of the Commission to Study the Organization of Peace (New York, 1969), pp. 24-25.

15. Commission to Study the Organization of Peace, "The United Nations and the Bed of the Sea II," Twenty-First Report of the Commission to Study the Organization of Peace (New York, 1970), pp. 41-42.

16. Report by the Special Subcommittee on the Outer Continental Shelf, "Outer Continental Shelf," 91st Congress, 2d session. This special subcommittee was chaired by Senator Metcalf, who noted that the report was accepted unanimously by members of the subcommittee without dissents or minority views. Congressional Record, March 10, 1971, p. 2814. The report was endorsed by Senator Henry Jackson, chairman of the full committee. See his comments in the Congressional Record of the same date, p. 2817.

17. Subcommittee Report, op. cit., p. 3.

18. Ibid., pp. 29-30.

19. Ibid., p. 5.

20. Ibid., p. 17.

21. Ibid.

22. On this point see Oliver J. Lissitzyn, International Law Today and Tomorrow (Dobbs Ferry: Oceana Publications, 1965), pp. 73-75.

23. Subcommittee Report, op. cit., p. 3.

24. This information is derived from an interview with a member of the staff of the Senate Interior Committee.

25. Subcommittee Report, op. cit., p. 25. See also "U.S. Alters Draft on Seabed Riches," New York Times, August 2, 1970, p. 27.

26. Senate Interior Committee, "Outer Continental Shelf," op. cit., part III, pp. 52-56.

27. The text of the letter to the NPC is reproduced in the Supplemental Report of the National Petroleum Council, op. cit., p. 33. The Supplemental Report was prepared in response to the request by Dole.

28. A/AC. 138/SR.29, p. 10.

29. See the comments of the representatives of Chile, A/AC. 138/SR.20, p. 14 and of the Soviet Union, A/AC. 138/SR.36, p. 94.

30. See, for example, the views of the representatives of Great Britain, A/AC. 138/SR.20, p. 20 and those of Australia, A/AC. 138/SR.34, p. 67.

31. A/AC. 138/SR.34, p. 64.

32. A/AC. 138/SR. 35, p. 72.

33. A/AC. 138/SR. 46, pp. 20-21.

34. A/AC. 138/SR. 49, p. 51.

35. United States Information Service, Press Release, Statement of John R. Stevenson, United States Representative to the Seabed Committee, August 18, 1971.

36. A copy was provided the author by the State Department.

37. A/AC. 138/SR.51, p. 76.

38. The text of this declaration appears in General Assembly Official Records, 27th session, supplement 21, A/8721, pp. 70-73.

39. Ibid., pp. 73-76.

40. See the testimony of Ambassador Stevenson before the House Merchant Marine and Fisheries Committee, Subcommittee on Oceanography, "Oceanography Miscellaneous," 92nd Congress, 2d session, p. 240.

41. A/AC. 138/SC. II/L.10. The text of this proposal also appears in A/8721, pp. 180-182.

42. A copy of this speech was provided the author by the State Department.

43. On this point see the testimony of John Moore, State Department Counselor on International Law, Senate Foreign Relations Committee, Subcommittee on Oceans and International Environment, "U.S. Oceans Policy," 93rd Congress, 1st session, p. 17.

44. General Assembly Official Records, Report of the Committee on the Peaceful Uses of the Sea-Bed and the Ocean Floor Beyond the Limits of National Jurisdiction, A/9021, volume III, pp. 19-22.

45. The text of this declaration is found in ibid., volume II, pp. 4-8.

46. Ibid., volume III, pp. 87-89.

47. Ibid., pp. 75-77.

48. Press Release of the U.S. Mission in Geneva, dated July 18, 1972, p. 2.

49. A/8721, pp. 175-179.

50. See the statement of Ambassador Stevenson to Subcommittee III of the UN Seabed Committee on August 2, 1972.

51. A/AC. 138/SC. III/L.40.

52. Ibid., article 15.

53. Ibid., article 17.

54. National Petroleum Council, Law of the Sea: Particular Aspects Affecting the Petroleum Industry (Washington, D.C., May 1973). See especially ch. 1. National Petroleum Council, Ocean Petroleum Resources: Interim Report of the National Petroleum Council (Washington, D. C. July 4, 1974) pp. 21-26.

55. Ibid., p. 10.

56. A/9021, volume II, pp. 39-165.

57. Ibid., pp. 52-53.

58. Ibid., pp. 57-58.

59. John R. Stevenson and Bernard Oxman, "The Preparations for the Law of the Sea Conference," American Journal of International Law 68 (1974): 6-7.

60. A/9021, volume II, p. 126.

61. Stevenson and Oxman, op. cit., p. 7.

62. A/AC. 138/SC. II/L.21, article 6 and A/AC. 138/SC. II/ L.40, article 7 respectively.

63. A/AC. 138/SC. III/L.44.

64. See the statement of Ambassador McKernan before Subcommittee III of the UN Seabed Committee on July 20, 1973.

65. A/AC. 138/SC. III/L.31.

66. A/AC. 138/SC. III/L.18, see especially point 9. See also the draft articles on scientific research of Brazil, Ecuador, El Salvador, Panama, Peru, and Uruguay, A/AC. 138/SC. III/L.45.

67. House Foreign Affairs Committee, Subcommittee on International Organizations and Movements, "Law of the Sea Resolution," 93rd Congress, 1st session, p. 46.

68. See the views expressed by Luke Finlay on behalf of the API, House Merchant Marine and Fisheries Committee, Subcommittee on Oceanography, "Oceanography Miscellaneous," 92nd Congress, 2d session, pp. 146-147. See also the July 1974 National Petroleum Council Report, op. cit. pp. 12-17. This report recommends the creation of an "economic resource zone" to extend to a distance of 200 miles from the baseline of the territorial sea or to the edge of the continental margin, whichever is broader.

69. Senate Foreign Relations Committee, Subcommittee on Oceans and International Environment, "U.S. Oceans Policy," 93rd Congress, 1st session, p. 2.

70. See, for example, the statement of John G. Laylin before the House Foreign Affairs Committee, Subcommittee on International

Organizations and Movements, "Law of the Sea and Peaceful Uses of the Seabed," 92nd Congress, 2d session, pp. 31-35, 103.

71. S.2801, H.R.13904, H.R.13076, H.R.14918 all introduced in 1972 and H.R.9 introduced in 1973.

72. See, for example, H.R.13076, section 4a.

73. House Committee on Merchant Marine and Fisheries, Subcommittee on Oceanography, "Deep Seabed Hard Mineral Resources," 92nd Congress, 2d session, pp. 9-16.

74. Testimony of John Flipse, President, Deepsea Ventures, Inc., ibid., p. 10.

75. Ibid., p. 98.

76. Ibid., p. 11.

77. See the study of F. L. LaQue, "Prospects For and From Deep Ocean Mining," in House Foreign Affairs Committee, "Law of the Sea and Peaceful Uses of the Seabed," op. cit., pp. 59-70.

78. House Committee on Merchant Marine and Fisheries, "Oceanography Miscellaneous," op. cit., p. 11.

79. See, for example, the statement made upon behalf of the American Mining Congress before the Senate Interior Committee, Subcommittee on Minerals, Materials, and Fuels, "Mineral Resources of the Deep Seabed," 93rd Congress, 1st session, pp. 178-200.

80. Senate Commerce Committee, "The Oceans and National Economic Development," 93rd Congress, 1st session, pp. 6-11.

81. Senate Interior Committee, Subcommittee on Minerals, Materials and Fuels, "Status Report on Law of the Sea Conference," 93rd Congress, 1st session, pp. 146-147.

82. See the May 31, 1972 letter of Edwin Hood, President of the Shipbuilders Council of America to the House Merchant Marine and Fisheries Committee, "Deep Seabed Hard Mineral Resources," 93rd Congress, 1st session, pp. 200-202.

83. Ibid., p. 45.

84. H.R.13076, section 10(a); H.R.9, section 10(a).

85. See, for example, the testimony of Professors Gary Knight, John Logue, and Wolfgang Friedmann before the House Merchant Marine and Fisheries Committee, "Oceanography Miscellaneous," op. cit., pp. 57-62, 149-167, and 167-186 respectively.

86. Ibid., p. 167.

87. Ibid., p. 156.

88. On this point see the May 19, 1972 letter of State Department Legal Adviser Stevenson to Senator Henry Jackson, reproduced in House Foreign Affairs Committee, "Law of the Sea and Peaceful Uses of the Seabeds," 92nd Congress, 2d session, pp. 24-25.

89. See the comments of John Stevenson on this matter before the Senate Interior Committee, "Status Report on the Law of the Sea Conference," op. cit., p. 11, and the study of the UN Seabed Committee on Provisional Application of Treaties, A/AC. 138/88.

90. Letter of Legal Adviser Brower to Senator Henry Jackson, dated March 1, 1973 reproduced in Senate Interior Committee, "Mineral Resources of the Deep Seabed," op. cit., pp. 15-23.

91. H.R.200, H.R.1576, 93rd Congress, 1st session. See also the Hearings before the Senate Commerce Committee, "Interim Fisheries Zone Extension and Management Act of 1973," 93rd Congress, 1st and 2d sessions.

92. See the report of the 1973 Northeast Fisheries Conference, New York Times, May 5, 1973, p. 45.

93. Public Law 93-242 signed into law on January 2, 1974, see especially section 15.

94. 24 U.S.T. 923.

95. House Report No. 688, 93rd Congress, 1st session, pp. 4-6.

96. Senate Report No. 633, 93rd Congress, 1st session, p. 7.

CHAPTER

8 CONCLUSIONS

With the development of technology and the increase in knowledge of the seas, the ocean areas of the world now open up to man a new frontier. Offshore lands, once of little or no importance, have today become significant and valuable. The continental shelf has emerged as a major new source of petroleum, and tremendous quantities of mineral resources are now recognized to exist in ocean bottom areas. Technological progress, which is making offshore resource exploitation possible, has had a military concomitant. Strategic weapons systems could be placed in seabed areas where their presence would be more difficult to detect, therefore making them less vulnerable to an enemy first strike. Sophisticated acoustical systems have, in fact, been placed in the seabed to track and thus make enemy submarines more vulnerable to destruction.

The continental shelf, however, has implications beyond its potential commercial and military uses. It is a key consideration in the future development of an international regime of the oceans generally. As seen, the United States government, from the time of the 1945 Truman Proclamation until very recently, has attempted to develop its policy with respect to ocean areas on the basis of a functional approach. The mineral resources of the continental shelf had thus been viewed as a matter apart from questions of fishery zones, territorial waters, or rights in epicontinental waters or airspace. This attempt at functional division of ocean space was an effort to serve simultaneously different American interests. The United States wanted the exclusive right to exploit the natural resources in its continental shelf; to this end it was willing to recognize that other states had this right with respect to their continental shelves. But while the United States would acquiesce to the claims of other states to mineral resources in their shelf areas, it opposed the assertions of rights to the living resources found in the waters above. American objections were clearly the consequence of American interests in distant water fisheries. Thus, according to the United States, different types of resources, even though they might be found at the same distance from the coast, were to be subject to different rules. Just as the issue of fishing rights was to be separated from the concept of the continental shelf, so too were the matters of territorial waters and the question of rights to free passage in the epicontinental seas and airspace. This distinction was also based in American interests, namely in maintaining maximum freedom of the seas. The claim contained in the Truman Proclamation, in light of these interests, was very limited in nature. It asserted for the United States jurisdiction and control over the mineral resources of the American continental shelf and claimed nothing more. The claim was tailor-made to suit then perceived United States needs and interests.

While a claim of this type made substantive sense in view of American interests, it did not similarly make sense from the perspective of a number of other states. For the Latin American states having a coast on the Pacific Ocean, a claim limited to the mineral resources of the continental shelf is virtually worthless, for there is little, if any, continental shelf to be found there. The distribution of continental shelf areas is, unfortunately, not at all uniform; this is a fact which has complicated international negotiations over seabed areas. If the American claim was substantively felt to be a meaningless precedent from the point of view of these Latin American states, it was suggestive, nevertheless, of a course of action more in line with perceived interests. If the United States could make unilateral claims in adjacent ocean areas for certain purposes, why could not other states do the same for other purposes? Since 1945 a number of states have made unilateral claims to ocean space surpassing and differing from that of the United States as embodied in the Truman Proclamation. It might thus be said that the United States, in seeking out its short term interest, made a significant mistake having long-term effect in asserting a unilateral claim to offshore mineral resources. Despite the fact that a number of states were asked for their views prior to the issuance of the Truman Proclamation, the major characteristic of that act was that it was unilateral in nature. Consequently, it opened the door for other states to make their own unilateral claims, some of which contained elements that were detrimental to important United States interests. In subsequent attempts to deny the validity of Latin American claims to wide ocean areas said to be subject to coastal state sovereignty or control, the United States has found that relations with these states have been adversely affected.

Aside from the resulting damage to American foreign relations and United States ocean interests, this experience also suggests that international law, from a world community perspective, is not best developed through unilateral actions. As Professor Henkin has observed, international law made by national assertion and reaction tends to increase the rights of states at the expense of international jurisdiction, whereas international law determined in a multilateral political forum is often an influence for restricting municipal interests in favor of broader interests. In brief, deliberations in international bodies may serve to encourage nations to recognize some principles of community interest.[1]

While the continental shelf is an issue of international politics and law, we have chosen to examine the development of American policy on this matter. We have attempted to demonstrate that this issue holds great importance for the United States and deserves the careful attention of American policy makers which of late it has been receiving. To understand American policy and motivations, we have delved into national politics as they are related to the continental shelf. We have considered the manner in which American international law positions on this question have been developed. We must conclude that they are the resultants of political pressures from within and without the government. American international law positions in this area have not represented the desires of some policy preference-free legally minded agency within the government such as the State Department Legal Adviser's Office, if indeed the Legal Adviser's Office could be characterized in such a manner. Certainly that office has played an important role in developing American policy, advising and cautioning on the consequences that might follow from different courses of action, and providing assistance in the drafting of proclamations and treaties. With the consideration by the United Nations of the matter of the seabeds, the Legal Adviser's Office took on the role of the "action office," becoming responsible for the advocacy of determined American policy. However, the Legal Adviser's Office must be viewed as only a small part of the total policy-making picture.

When we consider the shaping of the substantive positions favored by the United States, we find ourselves examining a complicated political situation in which a variety of interests seek to make themselves felt. We examined in detail the roles of the departments of Interior, State, and Defense and the various preferences and pressures that motivated their individual concerns and perspectives. We noted that in addition to these bureaucratic entities, there are a number of other parties that are actively seeking to influence American ocean policy; these interests include Congress, certain particular committees therein, the petroleum industry, the hard mineral industry, fishery interests, the scientific community, American nationalists, and American internationalists. Indeed, what we find is that there is growing recognition of the importance of ocean space and the activities that may occur there and this is reflected by the increasing number of actors who seek to affect American policy.

This study confirms the belief of Allison and Halperin that when a foreign policy problem arises each player in the intranational picture calculates how the resolution of the issue may affect his interests. Once the stakes are thus defined, stances on the particular issue are determined. In terms of bureaucratic politics, each participant observes a different face of the issue and thus it comes to pass that "one stands depending upon where one sits." Bureaucratic politics suggests that policy is made through what Allison refers to as a process of "pulling and hauling" rather than by grand design.[2] Decisions are thus resultants rather than chosen solutions to problems.

The model of bureaucratic politics is relevant to the study of determination of American international law positions. Considering the matter of the continental shelf, what we have seen is that within the United States government, until recently, it has been almost exclusively the Interior Department, which has determined American policy. It was the Interior Department that provided the original impetus resulting in the 1945 Truman Proclamation on the Continental Shelf. It was the Interior Department that was given day-to-day control over the shelf once the American claim had been made, and it was the Interior Department, through its leasing practices, that interpreted the ambiguous provisions of the 1958 Continental Shelf Convention for the United States. The Interior Department, like all branches of bureaucracy, examined developments and possibilities concerning offshore lands with a departmental perspective; there was seen to be a unity of interests between the Interior Department and the United States generally. James Rosenau has asserted that congressmen typically have "segmental" orientations rather than "continental" orientations. Segmentally oriented leaders give highest priority to the subnational units which they head or represent and are not prepared to admit a discrepancy between the national welfare and their subnational concerns.[3] What Rosenau has said of congressmen seems to hold as well for bureaucrats. In this connection it might be interesting to undertake a cross-national study of departmental interests and actions of the United States Department of the Interior and its foreign counterparts as a study of interest perception and even tacit cross-national functional alliances.

In terms of the daily interpretation of the Continental Shelf Convention, the Interior Department made the determination as to whether parcels of land desired for exploitation fell within the bounds of the "continental shelf." The State Department Legal Adviser's Office, when consulted, proved to be permissive in response to Interior's desires to lease more distant offshore lands. Believing as it did until at least the 1969-1970 period that a functional division of ocean space was possible, the State Department saw no need to object to the leasing policies of the Interior Department. In any case, it might well be argued that even had the State Department objected to Interior leasing in deeper and more distant waters, the Interior Department view would have prevailed in light of strong and influential constituency support behind that department. The State Department, not being a domestic operating agency with a powerful political clientele, has little political leverage, as seen, for example, in the annual congressional paring of State Department requests for foreign aid or embassy expense funds. When a forceful challenge to the Interior Department finally did emerge, it came from the Defense Department, as its analysts began to challenge the validity of a functional approach to ocean problems.

To return for a moment to the matter of constituency pressures, we find that the Interior Department position favoring a wide continental shelf reflected strong constituency preferences as well as departmental preferences. According to Peter Woll, the constituency of a bureaucratic agency consists of "those groups, both governmental and non-governmental, whose interests must be taken into account." Woll claims that because these interests must be taken into consideration, the agency by definition is responsible to its constituency.[4] Among the constituency groups thus defined, two of the most significant have been reinforcing the Interior Department's predilection toward a wide continental shelf, namely the large American oil companies and the Senate Interior Committee.

The large American oil companies have favored a wide continental shelf alternative and have done much to propagate their point of view. They have successfully urged that the Interior Department grant exploitation permits and leases in deeper and deeper waters off the coasts of the United States. They have vigorously defended the offshore leasing policies of the Interior Department before the Congress and the public as being a legitimate exercise of authority under the Outer Continental Shelf Lands Act and the Continental Shelf Convention. They strongly opposed the August 1970 American draft treaty and thus helped bring about its demise. According to a staff member of the Senate Foreign Relations Committee, representatives of the American petroleum industry appeared in Geneva during meetings of thc United Nations Seabed Committee and lobbied among foreign delegations for wide national continental shelf claims.

Despite industry disclaimers, it is difficult to believe that tax benefits and import quotas were not considered by the petroleum companies in reaching their conclusions in regard to ocean policy. Petroleum derived from an extended area of national jurisdiction offshore would not have been subject to any import quotas and American petroleum companies operating in similar areas off other states would clearly be entitled to foreign tax credits. On the other hand, the tax and import quota situation in the trusteeship zone and in the international zone beyond it as proposed by the August 1970 United States draft was not clear; in any case, it would not have been more favorable than if a wide shelf alternative were chosen. Furthermore, the oil companies would clearly prefer to continue doing business with national authorities rather than with an international body, because they have had extensive experience in dealings of this type. The operation of the United Nations with its majority of membership of less developed states has led to severe doubts in industry minds about the possibility of international jurisdiction and management. Thus, in petroleum industry circles, there is apparently no motive to favor change from the industry favored system of exclusive national jurisdiction over all the mineral resources of the continental margin.

As to the relevance of past experience in dealing with foreign governments for oil concessions, however, it might be suggested that such experience may not have that much bearing on the future. An examination of recent developments in international petroleum reveals that important changes have been recently occurring manifested, in large part, in the developing role of the Organization of Petroleum Exporting Countries (OPEC). With the development of a strong united front by oil exporting states, the traditional balance between oil companies from consuming states and the petroleum exporting states has been upset. As early as 1971 an American oil expert claimed that the balance had already shifted decisively in favor of the producing states.[5] Aside from merely forcing crude petroleum prices upward, OPEC has achieved acceptance for the demand of its membership for part and, eventually, majority share owning rights to the crude oil producing ventures in their territories.[6]

The charge of the oil industry that the energy needs of the United States were not taken into account by the August 1970 American draft treaty might be viewed as a subterfuge, since under the treaty the United States, in line with the concept of the trusteeship zone, would have had the power to determine when and by whom mineral exploitation could occur in the American continental margin. While it may be true that with the creation of the international regime then proposed by the United States some income from the continental margin would have been diverted to an international authority rather than to the United States Treasury, one must consider this relative to the quid pro quo: a stable, internationally agreed upon regime for ocean bottom development with significant guarantees for other important ocean uses.

Turning to the Senate Interior Committee, we note that its position seems to reflect that of the large oil companies, with which the committee has a clientele relationship, and reflects also a large measure of nationalism. The report of the Interior subcommittee points to the possible gains to the United States from a wide continental shelf, yet it nowhere considers the important questions raised by the Department of Defense. The report offers its interpretation of key clauses of the Continental Shelf Convention and then, in a statement which indicates its narrow approach to the whole question, observes:

> It appears to us that those who advocate a narrow Continental Shelf or a narrow Continental Shelf coupled with an intermediate zone, do not do so on legal grounds, that is on a reasoned interpretation of the legal doctrine of the Continental Shelf. Instead, it seems that their conclusions are grounded more on a policy preference.[7]

The tenor of this statement appears to favor reasoned legal interpretation as opposed to policy preferences in determining future American positions on the continental shelf. But is it not the case that policy preferences determine, to a very great extent, legal interpretation, especially where the law is unclear?

On the matter of continental shelf policy, the Defense Department and the Navy, in the past, have vigorously opposed the positions favored by the large petroleum companies. Unlike these companies and the Senate Interior Committee, the Defense Department and the Navy have not focused their attention upon provisions of the Continental Shelf Convention, but rather upon actual state practice.

As the State Department became involved in international negotiations on ocean matters in the General Assembly and later in the Seabed Committee, it gradually concluded that a piecemeal approach to ocean space would be inadequate. The various uses of ocean space were linked because of their effect upon one another and, perhaps more importantly, because other states perceived them to be related. The belief that the more encompassing views of some of the Latin American states were mere aberrations from the norm began to dissipate in light of the positions expressed by other less developed states. What finally emerged within the State Department was recognition of the need to deal with many ocean issues simultaneously and comprehensively.

To summarize, we find that beginning with the middle 1960s the matter of the seabeds acquired a more extensive interested public. With the Maltese initiative before the General Assembly, it became necessary for the United States government to develop some coherent point of view on seabed matters. Whereas in the past the Interior Department played the dominant role in relation to the continental shelf, the field of interested American governmental parties began to expand. In terms of Anthony Downs' concept of bureau territoriality,[8] the Interior Department began to see a challenge to its control in affairs relating to offshore areas. At least a twofold threat was involved. First, Interior was challenged by the Defense Department, whose differently perceived needs led to the desire for an oceans policy with an emphasis quite different from that favored by Interior. The Department of Defense, with its preferences based on arguments of national defense and security, and as a department which has great influence in Washington, was for Interior an especially strong opponent in the political process. Second, the Interior Department was disquieted by the fact that offshore lands were to be dealt with in some manner through the United Nations, where the United States, in the name of international diplomacy, might find itself making concessions to which the Interior Department would be opposed but would have little opportunity to prevent.

The Stratton Commission attempted to view offshore lands from a more disinterested vantage point than that of any of the specific competing interests, and sought to offer a policy alternative based on a more general view of United States concerns. It is thus not surprising that its suggestions involved a compromise of differing policy preferences. Subsequent American policy further reflected the attempt to please as many parties as possible while safeguarding perceived American military interests.

It is clear that the policy positions of the various intranational actors on continental shelf questions reflected the individual concerns and priorities of these actors. These policy preferences, in turn, seemed to imply the need for quite different courses of action in terms of the international law position of the United States. The resolution of policy differences and, consequently, those between different preferred international law positions occurred through the ongoing political process. Policy preferences and politics, rather than legal interpretations of treaties or custom, shaped American attitudes. The failure of the August 1970 draft treaty to win international support and the very significant opposition to it within the context of American politics led to its replacement by a more nationally minded proposal calling for the creation of an exclusive offshore seabed zone. Political support, both at home and abroad, is much more apparent for this new approach since it seems to embody a regime that is acceptable to a greater number of key interests. In the future order of the oceans it now seems evident that some variation of an economic zone will emerge as a major feature of any internationally agreed upon regime. Negotiations such as those undertaken in Caracas in 1974, if successful, will help to determine the specific nature of that zone.

Finally, this study suggests that international law must be seen as having intrastate as well as interstate effect. International law is thought of as influencing the actions of states vis-à-vis each other. It might also be seen as a means whereby a governmental department can protect its interests against undesired change by other actors within the same government. It might be suggested, for example, that the interpretation of the Continental Shelf Convention by the Interior Department was useful as a means of attempting to force other agencies within the American government to recognize an ever-growing offshore area as under the jurisdiction of the United States and, consequently, of the Interior Department. By allowing American operations to progress farther and farther offshore, in accordance with its interpretation of the Continental Shelf Convention, Interior was setting precedents in defining the exploitability clause. Other states could then take similar action in offshore areas adjacent to their own coasts. By thus encouraging this development, Interior would be making it more difficult for the Defense Department and other interests favoring a narrower area coastal state jurisdiction to attain their policy preferences. The proposed draft treaty of August 1970, in turn, served the basic interests of the Defense Department. Adoption of that treaty would have protected perceived Defense Department interests against undesired policy shifts favored by the Interior Department and oil interests. To bring about change in international law requires something more than a change in the preferences of American policy makers. Embodying certain values in international law may thus be seen as a means whereby some combination of domestic actors, to some extent, can lock their own government into a desired position.

NOTES

1. Louis Henkin, Law for the Seas Mineral Resources, op. cit., pp. 38, 49-50.

2. Graham Allison, Essence of Decision, op. cit., pp. 171, 175.

3. James Rosenau, National Leadership in Foreign Policy (Princeton: Princeton University Press, 1963), pp. 30-31, 347-350. Allison and Halperin, "Bureaucratic Politics," op. cit., p. 48, note that members of a department or agency come to believe that the health of their organization is vital to the national interest. This leads to the concern for the maintenance of autonomy and organizational morale, protection of the organization's essence, the maintenance or expansion of roles and missions, and the maintenance or increasing of budgets.

4. Peter Woll, American Bureaucracy (New York: W. W. Norton, 1963), pp. 53-54. See also Richard Neustadt, Presidential Power (New York: Signet Books, 1960), p. 47, and Aaron Wildavsky, The Politics of the Budgetary Process, op. cit., pp. 65-74.

5. Walter J. Levy, "Oil Power," *Foreign Affairs* 49 (1971): 652.

6. In two articles Fuad Itayim, editor of the *Middle East Economic Survey*, analyzes the agreement achieved in October 1972 between representatives of 23 international oil companies and Saudi Arabia, Iraq, Kuwait, Abu Dhadbi, and Qatar, New York *Times*, October 15, 1972, section 3, pp. 1, 3, and October 22, 1972, section 3, p. 1.

7. Interior Subcommittee Report, op. cit., p. 15.

8. Anthony Downs, *Inside Bureaucracy* (Boston: Little, Brown and Company, 1967), pp. 211-222. Downs notes that the "concept of *territoriality* has recently been developed in studies of animal behavior. These studies show that in many species of birds and animals, individuals, stake out and defend definite territories surrounding their nests or 'home bases.' The size of each territory depends upon the area required to generate an adequate food supply for its occupants, as well as the strength, skill and aggressiveness of those occupants. Some zoologists believe that territoriality is important for individual members of the species and acts as a device for controlling its total population in a given region. This concept is strikingly relevant to the behavior of bureaus." Pp. 212-213.

THE TRUMAN PROCLAMATION ON THE CONTINENTAL SHELF
(No. 2667, September 28, 1945)

Policy of the United States with Respect to the Natural Resources of the Subsoil and Seabed of the Continental Shelf
(By the President of the United States of America)

Whereas the Government of the United States of America, aware of the long range world-wide need for new sources of petroleum and other minerals, holds the view that efforts to discover and make available new supplies of these resources should be encouraged; and

Whereas its competent experts are of the opinion that such resources underlie many parts of the continental shelf off the coasts of the United States of America, and that with modern technological progress their utilization is already practicable or will become so at an early date; and

Whereas, recognized jurisdiction over these resources is required in the interest of their conservation and prudent utilization when and as development is undertaken; and

Whereas it is the view of the Government of the United States that the exercise of jurisdiction over the natural resources of the subsoil and seabed of the continental shelf by the contiguous nation is reasonable and just, since the effectiveness of measures to utilize or conserve these resources would be contingent upon cooperation and protection from the shore, since the continental shelf may be regarded as an extension of the land-mass of the coastal nation and thus naturally appurtenant to it, since these resources frequently form a seaward extension of a pool or deposit lying within the territory, and since self-protection compels the coastal nation to keep close watch over activities off its shores which are of the nature necessary for utilization of these resources;

Now, Therefore, I, Harry S. Truman, President of the United States of America, do hereby proclaim the following policy of the United States of America with respect to the natural resources of the subsoil and seabed of the continental shelf.

Having concern for the urgency of conserving and prudently utilizing its natural resources, the Government of the United States regards the natural resources of the subsoil and seabed of the continental shelf beneath the high seas but contiguous to the coasts of the United States as appertaining to the United States, subject to its jurisdiction and control. In cases where the continental shelf extends to the shores of another State, or is shared with an adjacent State, the boundary shall be determined by the United States and the State concerned in accordance with equitable principles. The character as high seas of the waters above the continental shelf and the right to their free and unimpeded navigation are in no way thus affected.

[SEAL] HARRY S. TRUMAN

By the President:
DEAN ACHESON,
Acting Secretary of State

APPENDIX B

THE TRUMAN PROCLAMATION ON FISHERIES
(No. 2668, September 28, 1945)

Policy of the United States with Respect to Coastal Fisheries in Certain Areas of the High Seas
(By the President of the United States of America)

Whereas for some years the Government of the United States of America has viewed with concern the inadequacy of present arrangements for the protection and perpetuation of the fishery resources contiguous to its coasts, and, in view of the potentially disturbing effect of this situation, has carefully studied the possibility of improving the jurisdictional basis for conservation measures and international cooperation in this field; and

Whereas such fishery resources have a special importance to coastal communities as a source of livelihood and to the nation as a food and industrial resource; and

Whereas the progressive development of new methods and techniques contributes to intensified fishing over wide sea areas and in certain cases seriously threatens fisheries with depletion; and

Whereas there is an urgent need to protect coastal fishery resources from destructive exploitation, having due regard to conditions peculiar to each region and situation and to the special rights and equities of the coastal State and of any other State which may have established a legitimate interest therein;

Now, therefore, I, Harry S. Truman, President of the United States of America, do hereby proclaim the following policy of the United States of America with respect to coastal fisheries in certain areas of the high seas:

In view of the pressing need for conservation and protection of fishery resources, the Government of the United States regards it as proper to establish conservation zones in those areas of the high seas contiguous to the coasts of the United States wherein fishing activities have been or in the future may be developed and maintained on a substantial scale. Where such activities have been or shall hereafter be developed and maintained by its nationals alone, the United States regards it as proper to establish explicitly bounded conservation zones in which fishing activities shall be subject to the regulation and control of the United States. Where such activities have been or shall hereafter be legitimately developed and maintained jointly by nationals of the United States and nationals of other States, explicitly bounded conservation zones may be established under agreements between the United States and such other States; and all fishing activities in such zones shall be subject to regulation and control as provided in such agreements. The right of any State to establish conservation zones off its shores in accordance with the above principles is conceded, provided that corresponding recognition is given to any fishing interests of nationals of the United States which may exist in such areas. The character as high seas of the areas in which such conservation zones are established and the right to their free and unimpeded navigation are in no way thus affected.

[SEAL] HARRY S. TRUMAN

By the President:
DEAN ACHESON,
Acting Secretary of State

WHITE HOUSE PRESS RELEASE OF SEPTEMBER 28, 1945

The Natural Resources of the Continental Shelf

The President today issued two proclamations asserting the jurisdiction of the United States over the natural resources of the continental shelf under the high seas contiguous to the coasts of the United States and its territories, and providing for the establishment of conservation zones for the protection of fisheries in certain areas of the high seas contiguous to the United States. The action of the President in regard to both the resources of the continental shelf and the conservation of high seas fisheries in which the United States has an interest was taken on the recommendation of the Secretary of State and the Secretary of the Interior.

Two companion Executive orders were also issued by the President. One reserved and set aside the resources of the continental shelf under the high seas and placed them for administrative purposes, pending legislative action, under the jurisdiction and control of the Secretary of the Interior. The other provided for the establishment by Executive orders, on recommendation of the Secretary of State and the Secretary of the Interior of fishery conservation zones in areas of the high seas contiguous to the coasts of the United States.

Until the present the only high seas fisheries in the regulation of which the United States has participated, under treaties or conventions are those for whales, Pacific halibut and fur seals.

In areas where fisheries have been or shall hereafter be developed and maintained by nationals of the United States alone, explicitly bounded zones will be set up in which the United States may regulate and control all fishing activities.

In other areas where the nationals of other countries as well as our own, have developed or shall hereafter legitimately develop fisheries, zones may be established by agreements between the United States and such other States and joint regulations and control will be put into effect.

The United States will recognize the rights of other countries to establish conservation zones off their own coasts where the interests of nationals of the United States are recognized in the same manner that we recognize the interests of the nationals of the other countries.

The assertion of this policy has long been advocated by conservationists, including a substantial section of the fishing industry of the United States, since regulation of a fishery resource within territorial waters cannot control the misuse or prevent the depletion of that resource through uncontrolled fishery activities conducted outside of the commonly accepted limits of territorial jurisdiction.

As a result of the establishment of this new policy, the United States will be able to protect effectively, for instance, its most valuable fishery, that for the Alaska salmon. Through painstaking conservation efforts and scientific management the United States has made excellent progress in maintaining the salmon at high levels. However, since the salmon spends a considerable portion of its life in the open sea, uncontrolled fishery activities on the high seas, either by nationals of the United States or other countries, have constituted an ever present menace to the salmon fishery.

The policy proclaimed by the President in regard to the jurisdiction over the continental shelf does not touch upon the question of Federal versus State control. It is concerned solely with establishing the jurisdiction of the United States from an international standpoint. It will, however, make possible the orderly development of an underwater area 750,000 square miles in extent. Generally, submerged land which is contiguous to the continent and which is covered by no more than 100 fathoms (600 feet) of water is considered as the continental shelf.

Petroleum geologists believe that portions of the continental shelf beyond the three-mile limit contain valuable oil deposits. The study of subsurface structures associated with oil deposits which have been discovered along the Gulf Coast of Texas, for instance, indicates that corresponding deposits may underlie the offshore or submerged land. The trend of oil-productive salt domes extends directly into the Gulf of Mexico off the Texas coast. Oil is also being taken at present from wells within the three-mile limit off the coast of California. It is quite possible, geologists say, that the oil deposits extend beyond this traditional limit of national jurisdiction.

Valuable deposits of minerals other than oil may also be expected to be found in these submerged areas. Ore mines now extend under the sea from the coasts of England, Chile, and other countries.

While asserting jurisdiction and control of the United States over the mineral resources of the continental shelf, the proclamation in no wise abridges the right of free and unimpeded navigation of waters of the character of high seas above the shelf, nor does it extend the present limits of the territorial waters of the United States.

The advance of technology prior to the present war had already made possible the exploitation of a limited amount of minerals from submerged lands within the three-mile limit. The rapid development of technical knowledge and equipment occasioned by the war, now makes possible the determination of the resources of the submerged lands outside of the three-mile limit. With the need for the discovery of additional resources of petroleum and other minerals it became advisable for the United States to make possible orderly development of these resources. The proclamation of the President is designed to serve this purpose.

APPENDIX D

THE SUBMERGED LANDS ACT (1953)

PUBLIC LAW 31

May 22, 1953
[H. R. 4198]

AN ACT To confirm and establish the titles of the States to lands beneath navigable waters within State boundaries and to the natural resources within such lands and waters, to provide for the use and control of said lands and resources, and to confirm the jurisdiction and control of the United States over the natural resources of the seabed of the continental shelf seaward of State boundaries

Submerged Lands Act

Be it enacted by the Senate and House of Representatives of the United States of America in Congress assembled, That this Act may be cited as the "Submerged Lands Act".

TITLE I

Definition

"Lands beneath navigable waters."

SEC. 2. When used in this Act—

(a) The term "lands beneath navigable waters" means—

(1) all lands within the boundaries of each of the respective States which are covered by nontidal waters that were navigable under the laws of the United States at the time such State became a member of the Union, or acquired sovereignty over such lands and waters thereafter, up to the ordinary high water mark as heretofore or hereafter modified by accretion, erosion, and reliction;

(2) all lands permanently or periodically covered by tidal waters up to but not above the line of mean high tide and seaward to a line three geographical miles distant from the coast line of each such State and to the boundary line of each such State where in any case such boundary as it existed at the time such State became a member of the Union, or as heretofore approved by Congress, extends seaward (or into the Gulf of Mexico) beyond three geographical miles, and

(3) all filled in, made, or reclaimed lands which formerly were lands beneath navigable waters, as hereinabove defined:

"Boundaries."

(b) The term "boundaries" includes the seaward boundaries of a State or its boundaries in the Gulf of Mexico or any of the Great Lakes as they existed at the time such State became a member of the Union, or as heretofore approved by the Congress, or as extended or confirmed pursuant to section 4 hereof but in no event shall the term "boundaries" or the term "lands beneath navigable waters" be interpreted as extending from the coast line more than three geographical miles into the Atlantic Ocean or the Pacific Ocean, or more than three marine leagues into the Gulf of Mexico;

"Coast line."

(c) The term "coast line" means the line of ordinary low water along that portion of the coast which is in direct contact with the open sea and the line marking the seaward limit of inland waters;

"Grantees" and "lessees."

(d) The terms "grantees" and "lessees" include (without limiting the generality thereof) all political subdivisions, municipalities, public and private corporations, and other persons holding grants or leases from a State, or from its predecessor sovereign if legally validated, to lands beneath navigable waters if such grants or leases were issued in accordance with the constitution, statutes, and decisions of the courts of the State in which such lands are situated, or of its predecessor sovereign: Provided, however, That nothing herein shall be construed as conferring upon said grantees or lessees any greater rights or interests other than are described herein and in their respective grants from the State, or its predecessor sovereign;

"Natural resources."

(e) The term "natural resources" includes, without limiting the generality thereof, oil, gas, and all other minerals, and fish, shrimp, oysters, clams, crabs, lobsters, sponges, kelp, and other marine animal and plant life but does not include water power, or the use of water for the production of power;

(f) The term "lands beneath navigable waters" does not include the beds of streams in lands now or heretofore constituting a part of the public lands of the United States if such streams were not meandered in connection with the public survey of such lands under the laws of the United States and if the title to the beds of such streams was lawfully patented or conveyed by the United States or any State to any person;

"Person."

(g) The term "State" means any State of the Union;

(h) The term "person" includes, in addition to a natural person, an association, a State, a political subdivision of a State, or a private, public, or municipal corporation.

TITLE II

Lands Beneath Navigable Waters Within State Boundaries

SEC. 3. Rights of the States

Title and powers.

(a) It is hereby determined and declared to be in the public interest that (1) title to and ownership of the lands beneath navigable waters within the boundaries of the respective States, and the natural resources within such lands and waters, and (2) the right and power to manage, administer, lease, develop, and use the said lands and natural resources all in accordance with applicable State law be, and they are hereby, subject to the provisions hereof, recognized, confirmed, established, and vested in and assigned to the respective States or the persons who were on June 5, 1950, entitled thereto under the law of the respective States in which the land is located, and the respective grantees, lessees, or successors in interest thereof;

Claims of U.S.

(b) (1) The United States hereby releases and relinquishes unto said States and persons aforesaid, except as otherwise reserved herein, all right, title, and interest of the United States, if any it has, in and to all said lands, improvements, and natural resources; (2) the United States hereby releases and relinquishes all claims of the United States, if any it has, for money or damages arising out of any operations of said States or persons pursuant to State authority upon or within said lands and navigable waters; and (3) the Secretary of the Interior or the Secretary of the Navy or the Treasurer of the United States shall pay to the respective States or their grantees issuing leases covering such lands or natural resources all moneys paid thereunder to the Secretary of the Interior or to the Secretary of the Navy or to the Treasurer of the United States and subject to the control of any of them or to the control of the United States on the effective date of this Act, except that portion of such moneys which (1) is required to be returned to a lessee; or (2) is deductible as provided by stipulation or agreement between the United States and any of said States;

Leases in effect on June 5, 1950

(c) The rights, powers, and titles hereby recognized, confirmed, established, and vested in and assigned to the respective States and their grantees are subject to each lease executed by a State, or its grantee, which was in force and effect on June 5, 1950, in accordance with its terms and provisions and the laws of the State issuing, or whose grantee issued, such lease, and such rights, powers, and titles are further subject to the rights herein now granted to any person holding any such lease to continue to maintain the lease, and to conduct operations thereunder, in accordance with its provisions, for the full term thereof, and any extensions, renewals, or replacements authorized therein, or heretofore authorized by the laws of the State issuing, or whose grantee issued such lease: Provided, however, That, if oil or gas was not being produced from such lease on and before December 11, 1950, or if the primary term of such lease has expired since December 11, 1950, then for a term from the effective date hereof equal to the term remaining unexpired on December 11, 1950, under the provisions of such lease or any extensions, renewals, or replacements authorized therein or heretofore authorized by the laws of the State issuing, or whose grantee issued, such lease: Provided, however, That within ninety days from the effective date hereof (i) the lessee shall pay to the State or its grantee issuing such lease all rents, royalties, and other sums payable between June 5, 1950, and the effective date hereof, under such lease and the laws of the State issuing or whose grantee issued such lease, except such rents, royalties, and other sums as have been paid to the State, its grantee, the Secretary of the Interior or the Secretary of the Navy or the Treasurer of the United States and not refunded to the lessee; and (ii) the lessee shall file with the Secretary of the Interior or the Secretary of the Navy and with the State issuing or whose grantee issued such lease, instruments consenting to the payment by the Secretary of the Interior or the Secretary of the Navy or the Treasurer of the United States to the State or its grantee issuing the lease, of all rents, royalties, and other payments under the control of the Secretary of the Interior or the Secretary of the Navy or the Treasurer of the United States or the United States which have been paid, under the lease, except such rentals, royalties, and other payments as have also been paid by the lessee to the State or its grantee;

Rights of U.S. respecting navigation, etc.

(d) Nothing in this Act shall affect the use, development, improvement, or control by or under the constitutional authority of the United States of said lands and waters for the purposes of navigation or flood control or the production of power, or be construed as the release or relinquishment of any rights of the United States arising under the constitutional authority of Congress to regulate or improve navigation, or to provide for flood control, or the production of power;

Surface waters west of 98th meridian.

(e) Nothing in this Act shall be construed as affecting or intended to affect or in any way interfere with or modify the laws of the States which lie wholly or in part westward of the ninety-eight meridian, relating to the ownership and control of ground and surface waters; and the control, appropriation, use, and distribution of such waters shall continue to be in accordance with the laws of such States.

SEC. 4. Seaward Boundaries—The seaward boundary of each original coastal State is hereby approved and confirmed as a line three geographical miles distant from its coast line or, in the case of the Great Lakes, to the international boundary. Any State admitted subsequent to the formation of the Union which has not already done so may extend its seaward boundaries to a line three geographical miles distant from its coast line, or to the international boundaries of the United States in the Great Lakes or any other body of water traversed by such boundaries. Any claim heretofore or herafter asserted either by constitutional provision, statute, or otherwise, indicating the intent of a State so to extend its boundaries is hereby approved and confirmed, without prejudice to its claim, if any it has, that its boundaries extend beyond that line. Nothing in this section is to be construed as questioning or in any manner prejudicing the existence of any State's seaward boundary beyond three geographical miles if it was so provided by its constitution or laws prior to or at the time such State became a member of the Union, or if it has been heretofore approved by Congress.

SEC. 5. Exceptions from Operation of Section 3 of this Act.—There is excepted from the operation of section 3 of this Act—

(a) all tracts or parcels of land together with all accretions thereto, resources therein, or improvements thereon, title to which has been lawfully and expressly acquired by the United States from any State or from any person in whom title had vested under the law of the State or of the United States, and all lands which the United States lawfully holds under the law of the State; all lands expressly retained by or ceded to the United States when the State entered the Union (otherwise than by a general retention or cession of lands underlying the marginal sea); all lands acquired by the United States by eminent domain proceedings, purchase, cession, gift, or otherwise in a proprietary capacity; all lands filled in, built up, or otherwise reclaimed by the United States for its own use; and any rights the United States has in lands presently and actually occupied by the United States under claim of right;

(b) such lands beneath navigable waters held, or any interest in which is held by the United States for the benefit of any tribe, band, or group of Indians or for individual Indians; and

(c) all structures and improvements constructed by the United States in the exercise of its navigational servitude.

SEC. 6. Powers Retained by the United States.—(a) The United States retains all its navigational servitude and rights in and powers of regulation and control of said lands and navigable waters for the constitutional purposes of commerce, navigation, national defense, and international affairs, all of which shall be paramount to, but shall not be deemed to include, proprietary rights of ownership, or the rights of management, administration, leasing, use, and development of the lands and natural resources which are specifically recognized, confirmed, established, and vested in and assigned to the respective States and others by section 3 of this Act.

(b) In time of war or when necessary for national defense, and the Congress or the President shall so prescribe, the United States shall have the right of first refusal to purchase at the prevailing market price, all or any portion of the said natural resources, or to acquire and use any portion of said lands by proceeding in accordance with due process of law and paying just compensation therefor.

SEC. 7. Nothing in this Act shall be deemed to amend, modify, or repeal the Acts of July 26, 1866 (14 Stat. 251), July 9, 1870 (16 Stat. 217), March 3, 1877 (19 Stat. 377), June 17, 1902 (32 Stat. 388), and December 22, 1944 (58 Stat. 887), and Acts amendatory thereof or supplementary thereto.

5 USC 485; 16 USC 460d, 825s; 30 USC 35, 36, 38, 43, 46, 47, 51, 52, 33 USC 701a-1 701c, 701f,

SEC. 8. Nothing contained in this Act shall affect such rights, if any, as may have been acquired under any law of the United States by any person in lands subject to this Act and such rights, if any, shall be governed by the law in effect at the time they may have

701j and notes 708, 709; 43 USC 321-323, 325, 327-329, 372-498 passim 661, 766.

been acquired: Provided, however, That nothing contained in this Act is intended or shall be construed as a finding, interpretation, or construction by the Congress that the law under which such rights may be claimed in fact or in law applies to the lands subject to this Act, or authorizes or compels the granting of such rights in such lands, and that the determination of the applicability or effect of such law shall be unaffected by anything contained in this Act.

Resources seaward of continental shelf.

SEC. 9. Nothing in this Act shall be deemed to affect in any wise the rights of the United States to the natural resources of that portion of the subsoil and seabed of the continental shelf lying seaward and outside of the area of lands beneath navigable waters, as defined in section 2 hereof, all of which natural resources appertain to the United States, and the jurisdiction and control of which by the United States is hereby confirmed.

SEC. 10. Executive Order Numbered 10426, dated January 16, 1953, entitled "Setting Aside Submerged Lands of the continental shelf as a Naval Petroleum Reserve", is hereby revoked insofar as it applies to any lands beneath navigable waters as defined in section 2 hereof.

18 FR 405.

SEC. 11. Separability—If any provision of this Act, or any section, subsection, sentence, clause, phrase or individual word, or the application thereof to any person or circumstance is held invalid, the validity of the remainder of the Act and of the application of any such provision, section, subsection, sentence, clause, phrase or individual word to other persons and circumstances shall not be affected thereby; without limiting the generality of the foregoing, if subsection 3 (a) 1, 3 (a) 2, 3 (b) 1, 3 (b) 2, 3 (b) 3, or 3, or 3 (c) or any provision of any of those subsections is held invalid, such subsection or provision shall be held separable and the remaining subsections and provisions shall not be affected thereby.

Approved May 22, 1953.

APPENDIX E

THE OUTER CONTINENTAL SHELF LANDS ACT (1953)

PUBLIC LAW 212

August 7, 1953
[H. R. 5134]

AN ACT To provide for the jurisdiction of the United States over the submerged lands of the outer continental shelf, and to authorize the Secretary of the Interior to lease such lands for certain purposes

Outer continental shelf Lands Act.

Be it enacted by the Senate and House of Representatives of the United States of America in in Congress assembled, That this Act may be cited as the "Outer Continental Shelf Lands Act".

SEC. 2. Definitions—When used in this Act—

(a) The term "outer continental shelf" means all submerged lands lying seaward and outside of the area of lands beneath navigable waters as defined in section 2 of the Submerged Lands Act (Public Law, 31, Eighty-third Congress, first session), and of which the subsoil and seabed appertain to the United States and are subject to its jurisdiction and control;

(b) The term "Secretary" means the Secretary of the Interior;

(c) The term "mineral lease" means any form of authorization for the exploration for, or development or removal of deposits of, oil, gas, or other minerals; and

(d) The term "person" includes, in addition to a natural person, an association, a State, a political subdivision of a State, or a private, public, or municipal corporation.

SEC. 3. Jurisdiction Over Outer Continental Shelf—(a) It is hereby declared to be the policy of the United States that the subsoil and seabed of the outer continental shelf appertain to the United States and are subject to its jurisdiction, control, and power of disposition as provided in this Act.

(b) This Act shall be construed in such manner that the character as high seas of the waters above the outer continental shelf and the right to navigation and fishing therein shall not be affected.

SEC. 4. Laws Applicable to Outer Continental Shelf—(a)

(1) The Constitution and laws and civil and political jurisdiction of the United States are hereby extended to the subsoil and seabed of the outer continental shelf and to all artificial islands and fixed structures which may be erected thereon for the purpose of exploring for, developing, removing, and transporting resources therefrom, to the same extent as if the outer continental shelf were an area of exclusive Federal jurisdiction located within a State: Provided, however, That mineral leases on the outer continental shelf shall be maintained or issued only under the provisions of this Act.

State laws.

(2) To the extent that they are applicable and not inconsistent with this Act or with other Federal laws and regulations of the Secretary now in effect or hereafter adopted, the civil and criminal laws of each adjacent State as of the effective date of this Act are hereby declared to be the law of the United States for that portion of the subsoil and seabed of the outer continental shelf, and artificial islands and fixed structures erected thereon, which would be within the area of the State if its boundaries were extended seaward to the outer margin of the outer continental shelf, and the President shall determine and publish in the Federal Register such projected lines extending

Publication of projected State lines.

seaward and defining each such area. All of such applicable laws shall be administered and enforced by the appropriate officers and courts of the United States. State taxation laws shall not apply to the outer continental shelf.

(3) The provisions of this section for adoption of State law as the law of the United States shall never be interpreted as a basis for claiming any interest in or jurisdiction on behalf of any State for any purpose over the seabed and subsoil of the outer continental shelf, or the property and natural resources thereof or the revenues therefrom.

Jurisdiction of U.S. district courts.

(b) The United States district courts shall have original jurisdiction of cases and controversies arising out of or in connection with any operations conducted on the outer continental shelf for the purpose of exploring for, developing, removing or transporting by pipeline the natural resources, or involving rights to the natural resources of the subsoil and seabed of the outer continental shelf, and proceedings with respect to any such case or controversy may be instituted in the judicial district in which any defendant resides or may be found, or in the judicial district of the adjacent State nearest the place where the cause of action arose.

Worker's compensation.

44 Stat. 1424.
33 USC 901.

(c) With respect to disability or death of an employee resulting from any injury occurring as the result of operations described in subsection (b), compensation shall be payable under the provisions of the Longshoremen's and Harbor Workers' Compensation Act. For the purposes of the extension of the provisions of the Longshoremen's and Harbor Workers' Compensation Act under this section—

(1) the term "employee" does not include a master or member of a crew of any vessel, or an officer or employee of the United States or any agency thereof or of any State or foreign government, or of any political subdivision thereof;

(2) the term "employer" means an employer any of whose employees are employed in such operations; and

(3) the term "United States" when used in a geographical sense includes the outer continental shelf and artificial islands and fixed structures thereon.

61 Stat. 136.
29 USC 167.

(d) For the purposes of the National Labor Relations Act, as amended, any unfair labor practice, as defined in such Act, occurring upon any artificial island or fixed structure referred to in subsection (a) shall be deemed to have occurred within the judicial district of the adjacent State nearest the place of location of such island or structure.

Coast Guard regulations, etc.

(e) (1) The head of the Department in which the Coast Guard is operating shall have authority to promulgate and enforce such reasonable regulations with respect to lights and other warning devices, safety equipment, and other matters relating to the promotion of safety of life and property on the islands and structures referred to in subsection (a) or on the waters adjacent thereto, as he may deem necessary.

(2) The head of the Department in which the Coast Guard is operating may mark for the protection of navigation any such island or structure whenever the owner has failed suitably to mark the same in accordance with regulations issued hereunder, and the owner shall pay the cost thereof.

Penalty.

Any person, firm, company, or corporation who shall fail or refuse to obey any of the lawful rules and regulations issued hereunder shall be quilty of a misdemeanor and shall be fined not more than $100 for each offense. Each day during which such violation shall continue shall be considered a new offense.

Artificial islands, etc.

(f) The authority of the Secretary of the Army to prevent obstruction to navigation in the navigable waters of the United States is hereby extended to artificial islands and fixed structures located on the outer continental shelf.

(g) The specific application by this section of certain provisions of law to the subsoil and seabed of the outer continental shelf and the artificial islands and fixed structures referred to in subsection (a) or to acts or offenses occurring or committed thereon shall not give rise to any inference that the application to such islands and structures, acts, or offenses of any other provision of law is not intended.

SEC. 5. Administration of Leasing of the Outer Continental Shelf—(a) (1) The Secretary shall administer the provisions of this Act relating to the leasing of the outer continental shelf, and shall prescribe such rules and regulations as may be necessary to carry out such provisions. The Secretary may at any time prescribe and amend such rules and regulations as he determines to be necessary and proper in order to provide for the prevention of waste and conservation of the natural resources of the outer continental shelf, and the protection of correlative rights therein, and, notwithstanding any other provisions herein, such rules and regulations shall apply to all operations conducted under a lease issued or maintained under the provisions of this Act. In the enforcement of conservation laws, rules, and regulations the Secretary is authorized to cooperate with the conservation agencies of the adjacent States. Without limiting the generality of the foregoing provisions of this section, the rules and regulations prescribed by the Secretary thereunder may provide for the assignment or relinquishment of leases, for the sale of royalty oil and gas accruing or reserved to the United States at not less than market value, and, in the interest of conservation, for unitization, pooling, drilling agreements, suspension of operations or production, reduction of rentals or royalties, compensatory royalty agreements,

Prevention of waste

Cooperation with State conservation agencies.

subsurface storage of oil or gas in any of said submerged lands, and drilling or other easements necessary for operations or production.

Penalty.

(2) Any person who knowingly and willfully violates any rule or regulation prescribed by the Secretary for the prevention of waste, the conservation of the natural resources, or the protection of correlative rights shall be deemed guilty of a misdemeanor and punishable by a fine of not more than $2,000 or by imprisonment for not more than six months, or by both such fine and imprisonment, and each day of violation shall be deemed to be a separate offense. The issuance and continuance in effect of any lease, or of any extension, renewal, or replacement of any lease under the provisions of the Act shall be conditioned upon compliance with the regulations issued under this Act and in force and effect on the date of the issuance of the lease if the lease is issued under the provisions of section 8 hereof, or with the regulations issued under the provisions of section 6(b), clause (2), hereof if the lease is maintained under the provisions of section 6 hereof.

Cancellation of lease.

(b) (1) Whenever the owner of a nonproducing lease fails to comply with any of the provisions of this Act, or the of lease, or of the regulations issued under this Act and in force and effect on the date of the issuance of the lease if the lease is issued under the provisions of section 8 hereof, or of the regulations issued under the provisions of section 6(b), clause (2), hereof, if the lease is maintained under the provisions of section 6 hereof, such lease may be canceled by the Secretary, subject to the right of judicial review as provided in section 8(j), if such default continues for the period of thirty days after mailing of notice by registered letter to the lease owner at his record post office address.

(2) Whenever the owner of any producing lease fails to comply with any of the provisions of this Act, or of the lease, or of the regulations issued under this Act and in force and effect on the date of the issuance of the lease if the lease is issued under the provisions of section 8 hereof, or of the regulations issued under the provisions of section 6(b), clause (2), hereof, if the lease is maintained under the provisions of section 6 hereof, such lease may be forfeited and canceled by an appropriate proceeding in any United States district court having jurisdiction under the provisions of section 4(b) of this Act.

Pipeline rights-of-way.

(c) Rights-of-way through the submerged lands of the outer continental shelf, whether or not such lands are included in a lease maintained or issued pursuant to this Act, may be granted by the Secretary for pipeline purposes for the transportation of oil, natural gas, sulphur, or other mineral under such regulations and upon such conditions as to the application therefor and the survey, location and width thereof as may be prescribed by the Secretary, and upon the express condition that such oil or gas pipelines shall transport or purchase without discrimination, oil or natural gas produced from said submerged lands in the vicinity of the pipeline in such proportionate amounts as the Federal Power Commission, in the case of gas, and the Interstate Commerce Commission, in the case of oil, may, after a full hearing with due notice thereof to the interested parties, determine to be reasonable, taking into account, among other things, conservation and the prevention of waste. Failure to comply with the provisions of this section or the regulations and conditions prescribed thereunder shall be ground for forfeiture of the grant in an appropriate judicial proceeding instituted by the United States in any United States district court having jurisdiction under the provisions of section 4 (b) of this Act.

Forfeiture of grant.

SEC. 6. Maintenance of Leases on Outer Continental Shelf—(a) The provisions of this section shall apply to any mineral lease covering submerged lands of the outer continental shelf issued by any State (including any extension, renewal, or replacement thereof heretofore granted pursuant to such lease or under the laws of such State) if—

Filing of lease, etc.

(1) such lease, or a true copy thereof, is filed with the Secretary by the lessee or his duly authorized agent within ninety days from the effective date of this Act, or within such period or periods as provided in section 7 hereof or as may be fixed from time to time by the Secretary;

(2) such lease was issued prior to December 21, 1948, and would have been on June 5, 1950, in force and effect in accordance with its terms and provisions and the law of the State issuing it had the State had authority to issue such lease;

(3) there is filed with the Secretary, within the period or periods specified in paragraph (1) of this subsection, (A) a certificate issued by the State official or agency having jurisdiction over such lease stating that it would have been in force and effect as required by the provisions of paragraph (2) of this subsection, or (B) in the absence of such certificate, evidence in the form of affidavits, receipts, canceled checks, or other documents that may be required by the Secretary, sufficient to prove that such lease would have been so in force and effect;

Sums payable.

(4) except as otherwise provided in section 7 hereof, all rents, royalties, and other sums payable under such lease between June 5, 1950, and the effective date of this Act, which have not been paid in accordance with the provisions thereof, or to the Secretary or to the

Secretary of the Navy, are paid to the Secretary within the period or periods specified in paragraph (1) of this subsection and all rents, royalties, and other sums payable under such lease after the effective date of this Act, are paid to the Secretary, who shall deposit such payments in the Treasury in accordance with section 9 of this Act;

(5) the holder of such lease certifies that such lease shall continue to be subject to the overriding royalty obligations existing on the effective date of this Act;

(6) such lease was not obtained by fraud or misrepresentation;

(7) such lease, if issued on or after June 23, 1947, was issued upon the basis of competitive bidding;

Royalty.

(8) such lease provides for a royalty to the lessor on oil and gas of not less than $12\frac{1}{2}$ per centum and on sulphur of not less than 5 per centum in amount or value of the production saved, removed, or sold from the lease, or, in any case in which the lease provides for a lesser royalty, the holder thereof consents in writing, filed with the Secretary, to the increase of the royalty to the minimum herein specified;

(9) the holder thereof pays to the Secretary within the period or periods specified in paragraph (1) of this subsection an amount equivalent to any severance, gross production, or occupation taxes imposed by the State issuing the lease on the production from the lease, less the State's royalty interest in such production, between June 5, 1950, and the effective date of this Act and not heretofore paid to the State, and thereafter pays to the Secretary as an additional royalty on the production from the lease, less the United States' royalty interest in such production, a sum of money equal

to the amount of the severance, gross production, or occupation taxes which would have been payable on such production to the State issuing the lease under its laws as they existed on the effective date of this Act;

Termination of lease.

(10) such lease will terminate within a period of not more than five years from the effective date of this Act in the absence of production or operations for drilling, or, in any case in which the lease provides for a longer period, the holder thereof consents in writing, filed with the Secretary, to the reduction of such period so that it will not exceed the maximum period herein specified; and

Surety bond.

(11) the holder of such lease furnishes such surety bond, if any, as the Secretary may require and complies with such other reasonable requirements as the Secretary may deem necessary to protect the interests of the United States.

Maintenance of lease.

(b) Any person holding a mineral lease, which as determined by the Secretary meets the requirements of subsection (a) of this section, may continue to maintain such lease, and may conduct operations thereunder, in accordance with (1) its provisions as to the area, the minerals covered, rentals and, subject to the provisions of paragraphs (8), (9), and (10) of subsection (a) of this section, as to royalties and as to the term thereof and of any extensions, renewals, or replacements authorized therein or heretofore authorized by the laws of the State issuing such lease, or, if oil or gas was not being produced in paying quantities from such lease on or before December 11, 1950, or if production in paying quantities has ceased since June 5, 1950, or if the primary term of such lease has expired since December 11, 1950, then for a term from the effective date hereof equal to the term remaining unexpired on December 11, 1950, under the provisions of such lease or any extensions, renewals, or replacements authorized therein, or heretofore authorized by the laws of such State, and (2) such regulations as the Secretary may under section 5 of this Act prescribe within ninety days after making his determination that such lease meets the requirements of subsection (a) of this section: Provided, however, That any rights to sulphur under any lease maintained under the provisions of this subsection shall not extend beyond the primary term of such lease or any extension thereof under the provisions of such subsection (b) unless sulphur is being produced in paying quantities or drilling, well reworking, plant construction, or other operations for the production of sulphur, as approved by the Secretary, are being conducted on the area covered by such lease on the date of expiration of such primary term or extension: Provided further, That if sulphur is being produced in paying quantities on such date, then such rights shall continue to be maintained in accordance with such lease and the provisions of this Act: Provided further, That, if the primary term of a lease being maintained under subsection (b) hereof has expired prior to the effective date of this Act and oil or gas is being produced in paying quantities on such date, then such rights to sulphur as the lessee may have under such lease shall continue for twenty-four months from the effective date of this Act and as long thereafter as sulphur is produced in paying quantities, or drilling, well working, plant construction, or other operations for the production of sulphur, as approved by the Secretary, are being conducted on the area covered by the lease.

Sulphur.

Nonwaiver of U.S. claims.

(c) The permission granted in subsection (b) of this section shall not be construed to be a waiver of such claims, if any, as the United States may have against the lessor or the lessee or any other person respecting sums payable or paid for or under the lease, or respecting activities conducted under the lease, prior to the effective date of this Act.

Court review of determination.

(d) Any person complaining of a negative determination by the Secretary of the Interior under this section may have such determination reviewed by the United States District Court for the District of Columbia by filing a petition for review within sixty days after receiving notice of such action by the Secretary.

Lands beneath navigable waters.

(e) In the event any lease maintained under this section covers lands beneath navigable waters, as that term is used in the Submerged Lands Act, as well as lands of the outer continental shelf, the provisions of this section shall apply to such lease only insofar as it covers lands of the outer continental shelf.

Agreements with State.

SEC. 7. Controversy Over Jurisdiction— In the event of a controversy between the United States and a State as to whether or not lands are subject to the provisions of this Act, the Secretary is authorized, notwithstanding the provisions of subsections (a) and (b) of section 6 of this Act, and with the concurrence of the Attorney General of the United States, to negotiate and enter into agreements with the State, its political subdivision or grantee or a lessee thereof, respecting operations under existing mineral leases and payment and impounding of rents, royalties, and other sums payable thereunder, or with the State, its political subdivision or grantee, respecting the issuance or non-issuance of new mineral leases pending the settlement or adjudication of the controversy. The authorization contained in the preceding sentence of this section shall not be construed to be a limitation upon the authority conferred on the Secretary in other sections of this Act. Payments made pursuant to such agreement, or pursuant to any stipulation between the United States and a State, shall be considered as compliance with section 6 (a) (4) hereof. Upon the termination of such agreement or stipulation by reason of the final settlement or adjudication of such controversy, if the lands subject to any mineral lease are determined to be in whole or in part lands subject to the provisions of this Act, the lessee, if he has not already done so, shall comply with the requirements of section 6(a), and thereupon the provisions of section 6 (b) shall govern such lease. The notice concerning "Oil and Gas Operations in the Submerged Coastal Lands of the Gulf of Mexico" issued by the Secretary on December 11, 1950 (15 F. R. 8835), as amended by the notice dated January 26, 1951 (16 F. R. 953), and as supplemented by the notices dated February 2, 1951 (16 F. R. 1203), March 5, 1951 (16 F. R. 2195), April 23, 1951 (16 F. R. 3623), June 25, 1951 (16 F. R. 6404), August 22, 1951 (16 F. R. 8720), October 24, 1951 (16 F. R. 10998), December 21, 1951 (17 F. R. 43), March 25, 1952 (17 F. R. 2821), June 26, 1952 (17 F. R. 5833), and December 24, 1952 (18 F. R. 48), respectively, is hereby approved and confirmed.

Bids. Oil and gas leases.

SEC. 8. Leasing of Outer Continental Shelf—(a) In order to meet the urgent need for further exploration and development of the oil and gas deposits of the submerged lands of the outer continental shelf, the Secretary is authorized to grant to the highest responsible qualified bidder by competitive bidding under regulations promulgated in advance, oil and gas leases on submerged lands of the outer continental shelf which are not covered by leases meeting the requirements of subsection (a) of section 6 of this Act. The bidding shall be (1) by sealed bids, and (2) at the discretion of the Secretary, on the basis of a cash bonus with a royalty fixed by the Secretary at not less than $12\frac{1}{2}$ per centum in amount or value of the production saved, removed or sold, or on the basis of royalty, but at not less than the per centum above mentioned, with a cash bonus fixed by the Secretary.

(b) An oil and gas lease issued by the Secretary pursuant to this section shall (1) cover a compact area not exceeding five thousand seven hundred and sixty acres, as the Secretary may determine, (2) be for a period of five years and as long thereafter as oil or gas may be produced from the area in paying quantities, or drilling or well reworking operations as approved by the Secretary are conducted thereon, (3) require the payment of a royalty of not less than $12\frac{1}{2}$ per centum, in the amount or value of the production saved, removed, or sold from the lease, and (4) contain such rental provisions and such other terms and provisions as the Secretary may prescribe at the time of offering the area for lease.

Sulphur leases.

(c) In order to meet the urgent need for further exploration and development of the sulphur deposits in the submerged lands of the outer continental shelf, the Secretary is authorized to grant to the qualified persons offering the highest cash bonuses on a basis of competitive bidding sulphur leases on submerged lands of the outer continental shelf, which are not covered by leases which include sulphur and meet the requirements of subsection (a) of section 6 of this Act, and which sulphur leases shall be offered for bid by sealed bids and granted on separate leases from oil and gas leases, and for a separate consideration, and without priority or preference accorded to oil and gas lessees on the same area.

(d) A sulphur lease issued by the Secretary pursuant to this section shall (1) cover an area of such size and dimensions as the Secretary may determine, (2) be for a period of not more than ten years and so long thereafter as sulphur may be produced from the area in paying quantities or drilling, well reworking, plant construction, or other operations for the production of sulphur, as approved by the Secretary, are conducted thereon, (3) require the payment to the United States of such royalty as may be specified in the lease but not less than 5 per centum of the gross production or value of the sulphur at the wellhead, and (4) contain such rental provisions and such other terms and provisions as the Secretary may by regulation prescribe at the time of offering the area for lease.

Other mineral leases.

(e) The Secretary is authorized to grant to the qualified persons offering the highest cash bonuses on a basis of competitive bidding leases of any mineral other than oil, gas, and sulphur in any area of the outer continental shelf not then under lease for such mineral upon such royalty, rental, and other terms and conditions as the Secretary may prescribe at the time of offering the area for lease.

Notices, publication.

(f) Notice of sale of leases, and the terms of bidding, authorized by this section shall be published at least thirty days before the date of sale in accordance with rules and regulations promulgated by the Secretary.

Deposits.

(g) All moneys paid to the Secretary for or under leases granted pursuant to this section shall be deposited in the Treasury in accordance with section 9 of this Act.

(h) The issuance of any lease by the Secretary pursuant to this Act, or the making of any interim arrangements by the Secretary pursuant to section 7 of this Act shall not prejudice the ultimate settlement or adjudication of the question as to whether or not the area involved is in the outer continental shelf.

Cancellation.

(i) The Secretary may cancel any lease obtained by fraud or misrepresentation.

(j) Any person complaining of a cancellation of a lease by the Secretary may have the Secretary's action reviewed in the United States District Court for the District of Columbia by filing a petition for review within sixty days after the Secretary takes such action.

SEC. 9. Disposition of Revenues—All rentals, royalties, and other sums paid to the Secretary or the Secretary of the Navy under any lease on the outer continental shelf for the period from June 5, 1950, to date, and thereafter shall be deposited in the Treasury of the United States and credited to miscellaneous receipts.

SEC. 10. Refunds—(a) Subject to the provisions of subsection (b) hereof, when it appears to the satisfaction of the Secretary that any person has made a payment to the United States in connection with any lease under this Act in excess of the amount he was lawfully required to pay, such excess shall be repaid without interest to such person or his legal representative, if a request for repayment of such excess is filed with the Secretary within two years after the making of the payment, or within ninety days after the effective date of this Act. The Secretary shall certify the amounts of all such repayments to the Secretary of the Treasury, who is authorized and directed to make such repayments out of any money in the special account established under section 9 of this Act and to issue his warrant in settlement thereof.

Report to Congress.

(b) No refund of or credit for such excess payment shall be made until after the expiration of thirty days from the date upon which a report giving the name of the person to whom the refund or credit is to be made, the amount of such refund or credit, and a summary of the facts upon which the determination of the Secretary was made is submitted to the President of the Senate and the Speaker of the House of Representatives for transmittal to the appropriate legislative committee of each body, respectively; Provided, That if the Congress shall not be in session on the date of such submission or shall adjourn prior to the expiration of thirty days from the date of such submission, then such payment or credit shall not be made until thirty days after the opening day of the next succeeding session of Congress

SEC. 11. Geological and Geophysical Explorations—Any agency of the United States and any person authorized by the Secretary may conduct geological and geophysical explorations in the outer continental shelf, which do not interfere with or endanger actual operations under any lease maintained or granted pursuant to this Act, and which are not unduly harmful to aquatic life in such area.

SEC. 12. Reservations—(a) The President of the United States may, from time to time, withdraw from disposition any of the unleased lands of the outer continental shelf.

War.

(b) In time of war, or when the President shall so prescribe, the United States shall have the right of first refusal to purchase at the market price all or any portion of any mineral produced from the outer continental shelf.

(c) All leases issued under this Act, and leases, the maintenance and operation of which are authorized under this Act, shall contain or be construed to contain a provision whereby authority is vested in the Secretary, upon a recommendation of the Secretary of Defense, during a state of war or national emergency declared by the Congress or the President of the United States after the effective date of this Act, to suspend operations under any lease; and all such leases shall contain or be construed to contain provisions for the payment of just compensation to the lessee whose operations are thus suspended.

National defense areas.

(d) The United States reserves and retains the right to designate by and through the Secretary of Defense, with the approval of the President, as areas restricted from exploration and operation that part of the outer continental shelf needed for national defense; and so long as such designation remains in effect no exploration or operations may be conducted on any part of the surface of such area except with the concurrence of the Secretary of Defense; and if operations or production under any lease theretofore issued on lands within any such restricted area shall be suspended, and payment of rentals, minimum royalty, and royalty prescribed by such lease likewise shall be suspended during such period of suspension of operation and production, and the term of such lease shall be extended by adding thereto any such suspension period, and the United States shall be liable to the leasee for such compensation as is required to be paid under the Constitution of the United States.

Uranium, thorium, etc. 60 Stat. 760 42 USC 1805.

(e) All uranium, thorium, and all other materials determined pursuant to paragraph (1) of subsection (b) of section 5 of the Atomic Energy Act of 1946, as amended, to be peculiarly essential to the production of fissionable material, contained, in whatever concentration, in deposits in the subsoil or seabed of the outer continental shelf are hereby reserved for the use of the United States.

Helium.

(f) The United States reserves and retains the ownership of and the right to extract all helium, under such rules and regulations as shall be prescribed by the Secretary, contained in gas produced from any portion of the outer continental shelf which may be subject to any lease maintained or granted pursuant to this Act, but the helium shall be extracted from such gas as to cause no substantial delay in the delivery of gas produced to the purchaser of such gas.

18 F. R. 405.

SEC. 13. Naval Petroleum Reserve Executive Order Repealed—Executive Order Numbered 10426, dated January 16, 1953, entitled "Setting Aside Submerged Lands of the continental shelf as a Naval Petroleum Reserve", is hereby revoked.

SEC. 14. Prior Claims Not Affected—Nothing herein contained shall affect such rights, if any, as may have been acquired under any law of the United States by any person in lands subject to this Act and such rights, if any, shall be governed by the law in effect at the time they may have been acquired: Provided, however, That nothing herein contained is intended or shall be construed as a finding, interpretation, or construction by the Congress that the law under which such rights may be claimed in fact applies to the lands subject to this Act or authorizes or compels the granting of such rights in such lands, and that the determination of the applicability or effect of such law shall be unaffected by anything herein contained.

SEC. 15. Report by Secretary—As soon as practicable after the end of each fiscal year, the Secretary shall submit to the President of the Senate and the Speaker of the House of Representatives a report detailing the amounts of all moneys received and expended in connection with the administration of this Act during the preceding fiscal year.

SEC. 16. Appropriations—There is hereby authorized to be appropriated such sums as may be necessary to carry out the provisions of this Act.

SEC. 17. Separability—If any provision of this Act, or any section, subsection, sentence, clause, phrase or individual word, or the application thereof to any person or circumstance is held invalid, the validity of the remainder of the Act and of the application of any such provision, section, subsection, sentence, clause, phrase or individual word to other persons and circumstances shall not be affected thereby.

Approved August 7, 1953.

APPENDIX F

THE CONTINENTAL SHELF CONVENTION OF 1958

The States Parties to this Convention have agreed as follows:

Article 1

For the purpose of these articles, the term "continental shelf" is used as referring (a) to the seabed and subsoil of the submarine areas adjacent to the coast but outside the area of the territorial sea, to a depth of 200 metres or, beyond that limit, to where the depth of the superjacent waters admits of the exploitation of the natural resources of the said areas; (b) to the seabed and subsoil of similar submarine areas adjacent to the coasts of islands.

Article 2

1. The coastal State exercises over the continental shelf sovereign rights for the purpose of exploring it and exploiting its natural resources.

2. The rights referred to in paragraph 1 of this article are exclusive in the sense that if the coastal State does not explore the continental shelf or exploit its natural resources, no one may undertake these activities, or make a claim to the continental shelf, without the express consent of the coastal State.

3. The rights of the coastal State over the continental shelf do not depend on occupation, effective or notional, or on any express proclamation.

4. The natural resources referred to in these articles consist of the mineral and other non-living resources of the seabed and subsoil together with living organisms belonging to sedentary species, that is to say, organisms which, at the harvestable stage, either are immobile on or under the seabed or are unable to move except in constant physical contact with the seabed or the subsoil.

Article 3

The rights of the coastal State over the continental shelf do not affect the legal status of the superjacent waters as high seas, or that of the airspace above those waters.

Article 4

Subject to its right to take reasonable measures for the exploration of the continental shelf and the exploitation of its natural resources, the coastal State may not impede the laying or maintenance of submarine cables or pipe lines on the continental shelf.

Article 5

1. The exploration of the continental shelf and the exploitation of its natural resources must not result in any unjustifiable interference with navigation, fishing or the conservation of the living resources of the sea, nor result in any interference with fundamental oceanographic or other scientific research carried out with the intention of open publication.

2. Subject to the provisions of paragraphs 1 and 6 of this article, the coastal State is entitled to construct and maintain or operate on the continental shelf installations and other devices necessary for its exploration and the exploitation of its natural resources, and to establish safety zones around such installations and devices and to take in those zones measures necessary for their protection.

3. The safety zones referred to in paragraph 2 of this article may extend to a distance of 500 metres around the installations and other devices which have been erected, measured from each point of their outer edge. Ships of all nationalities must respect these safety zones.

4. Such installations and devices, though under the jurisdiction of the coastal State, do not possess the status of islands. They have no territorial sea of their own, and their presence does not affect the delimitation of the territorial sea of the coastal State.

5. Due notice must be given of the construction of any such installations, and permanent means for giving warning of their presence must be maintained. Any installations which are abandoned or disused must be entirely removed.

6. Neither the installations or devices, nor the safety zones around them, may be established where interference may be caused to the use of recognized sea lanes essential to international navigation.

7. The coastal State is obliged to undertake, in the safety zones, all appropriate measures for the protection of the living resources of the sea from harmful agents.

8. The consent of the coastal State shall be obtained in respect of any research concerning the continental shelf and undertaken there. Nevertheless the coastal State shall not normally withold its consent if the request is submitted by a qualified institution with a view to purely scientific research into the physical or biological characteristics of the continental shelf, subject to the proviso that the coastal State shall have the right, if it so desires, to participate or to be represented in the research, and that in any event the results shall be published.

Article 6

1. Where the same continental shelf is adjacent to the territories of two or more States whose coasts are opposite each other, the boundary of the continental shelf appertaining to such States shall be determined by agreement between them. In the absence of agreement, and unless another boundary line is justified by special circumstances, the boundary is the median line, every point of which is equidistant from the nearest points of the baselines from which the breadth of the territorial sea of each State is measured.

2. Where the same continental shelf is adjacent to the territories of two adjacent States, the boundary of the continental shelf shall be determined by agreement between them. In the absence of agreement, and unless another boundary line is justified by special circumstances, the boundary shall be determined by application of the principle of equidistance from the nearest points of the baselines from which the breadth of the territorial sea of each State is measured.

3. In delimiting the boundaries of the continental shelf, any lines which are drawn in accordance with the principles set out in paragraphs 1 and 2 of this article should be defined with reference to charts and geographical features as they exist at a particular date, and reference should be made to fixed permanent identifiable points on the land.

Article 7

The provisions of these articles shall not prejudice the right of the coastal State to exploit the subsoil by means of tunnelling irrespective of the depth of water above the subsoil.

Article 8

This Convention shall, until 31 October 1958, be open for signature by all States Members of the United Nations or of any of the specialized agencies, and by any other State invited by the General Assembly of the United Nations to become a Party to the Convention.

Article 9

This Convention is subject to ratification. The instruments of ratification shall be deposited with the Secretary-General of the United Nations.

Article 10

This Convention shall be open for accession by any States belonging to any of the categories mentioned in article 8. The instruments of accession shall be deposited with the Secretary-General of the United Nations.

Article 11

1. This Convention shall come into force on the thirtieth day following the date of deposit of the twenty-second instrument of ratification or accession with the Secretary-General of the United Nations.
2. For each State ratifying or acceding to the Convention after the deposit of the twenty-second instrument of ratification or accession, the Convention shall enter into force on the thirtieth day after deposit by such State of its instrument of ratification or accession.

Article 12

1. At the time of signature, ratification or accession, any State may make reservations to articles of the Convention other than to articles 1 to 3 inclusive.
2. Any Contracting State making a reservation in accordance with the preceding paragraph may at any time withdraw the reservation by a communication to that effect addressed to the Secretary-General of the United Nations.

Article 13

1. After the expiration of a period of five years from the date on which this Convention shall enter into force, a request for the revision of this Convention may be made at any time by any Contracting Party by means of a notification in writing addressed to the Secretary-General of the United Nations.

2. The General Assembly of the United Nations shall decide upon the steps, if any, to be taken in respect of such request.

Article 14

The Secretary-General of the United Nations shall inform all States Members of the United Nations and the other States referred to in article 8:

(a) Of signatures to this Convention and of the deposit of instruments of ratification or accession, in accordance with article 8, 9 and 10;
(b) Of the date on which this Convention will come into force, in accordance with article 11;
(c) Of requests for revision in accordance with article 13;
(d) Of reservations to this Convention, in accordance with article 12.

Article 15

The original of this Convention, of which the Chinese, English, French, Russian and Spanish texts are equally authentic, shall be deposited with the Secretary-General of the United Nations, who shall send certified copies thereof to all States referred to in article 8.

APPENDIX G

DIAGRAMMATIC PROFILE OF THE CONTINENTAL MARGIN

Average Widths and Depths

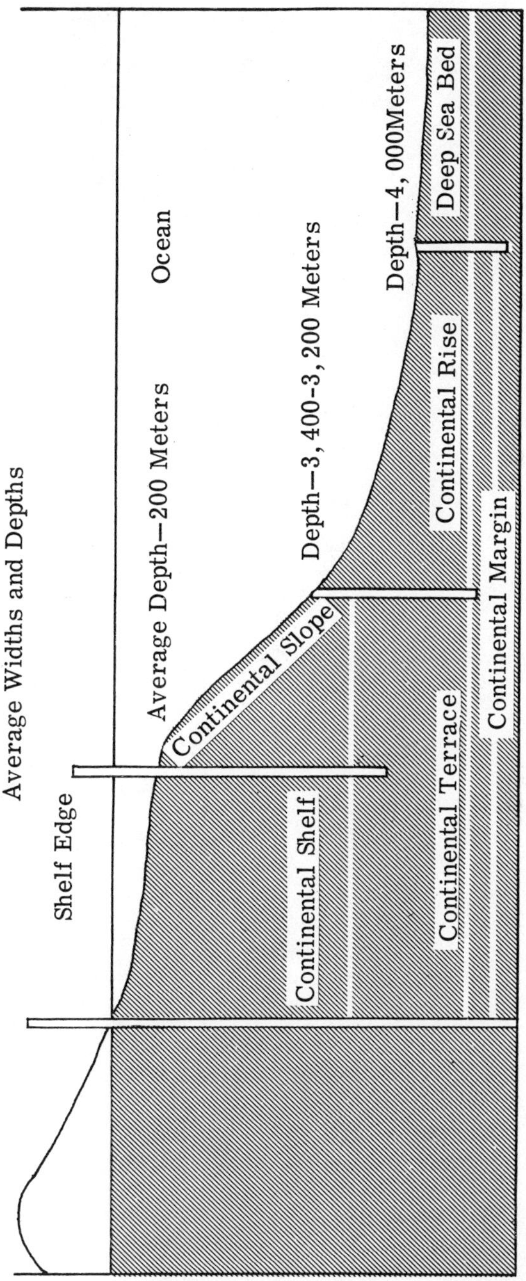

Profiles of U. S. Continental Margins

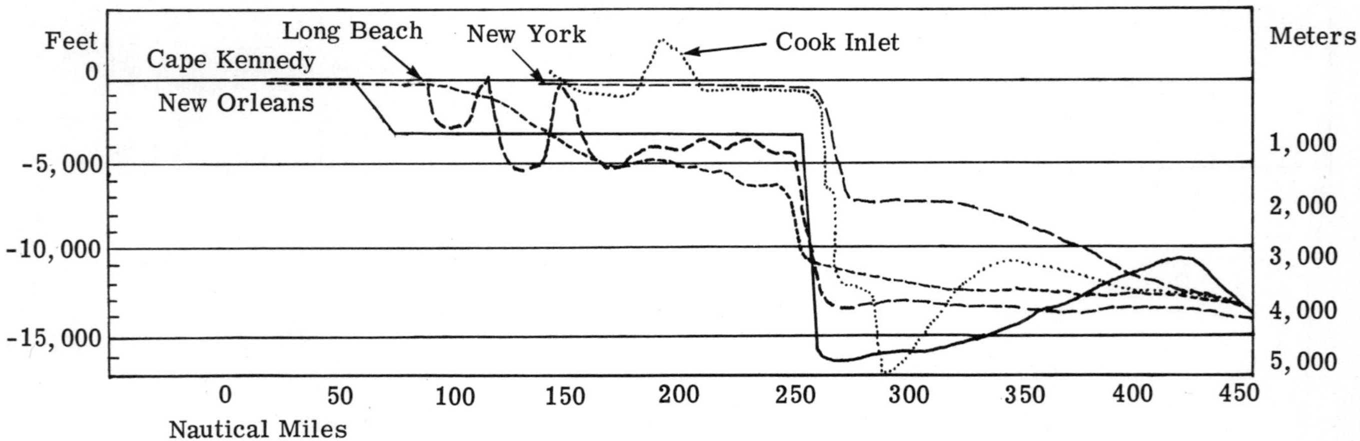

Source: Senate Interior Committee, Special Subcommittee on the Outer Continental Shelf, "Report of the Special Subcommittee on the Outer Continental Shelf," 91st Congress, 2d session, p. 222.

APPENDIX H

UN GENERAL ASSEMBLY RESOLUTION 2574 D (XXIV), THE "MORATORIUM" RESOLUTION, AND THE VOTE ON IT (1969)

The General Assembly,

Recalling its resolution 2467 A (XXIII) of 21 December 1968 to the effect that the exploitation of the resources of the seabed and the ocean floor, and the subsoil thereof, beyond the limits of national jurisdiction, should be carried out for the benefit of mankind as a whole, irrespective of the geographical location of States, taking into account the special interests and needs of the developing countries,

Convinced that it is essential, for the achievement of this purpose, that such activities be carried out under an international regime, including appropriate international machinery,

Noting that this matter is under consideration by the Committee on the Peaceful Uses of the Sea-Bed and the Ocean Floor beyond the Limits of National Jurisdiction,

Recalling its resolution 2340 (XXII) of 18 December 1967 on the importance of preserving the seabed and the ocean floor, and the subsoil thereof, beyond the limits of national jurisdiction, from actions and uses which might be detrimental to the common interests of mankind,

Declares that, pending the establishment of the aforementioned international régime:

(a) States and persons, physical or juridical, are bound to refrain from all activities of exploitation of the resources of the area of the seabed and ocean floor, and the subsoil thereof, beyond the limits of national jurisdiction:

(b) No claim to any part of that area or its resources shall be recognized.

1833rd plenary meeting,
15 December 1969.

Recorded Vote on Resolution D

In favour.—Afghanistan, Algeria, Argentina, Barbados, Bolivia, Brazil, Burundi, Central African Republic, Ceylon, Chad, Chile, Colombia, Congo (Brazzaville), Congo (Democratic Republic of), Costa

Rica, Cyprus, Dahomey, Dominican Republic, Ecuador, Ethiopia, Finland, Guatemala, Guinea, Guyana, Haiti, Honduras, India, Iraq, Jamaica, Jordan, Kenya, Kuwait, Lesotho, Malaysia, Maldives, Mali, Mauritania, Mauritius, Mexico, Morocco, Nepal, Nicaragua, Niger, Pakistan, Panama, Paraguay, Peru, Rwanda, Singapore, Somalia, Southern Yemen, Sweden, Thailand, Trinidad and Tobago, Tunisia, Uganda, United Republic of Tanzania, Uruguay, Venezuela, Yemen, Yugoslavia, Zambia.

Against.—Australia, Austria, Belgium, Bulgaria, Byelorussia, Canada, Czechoslovakia, Denmark, France, Ghana, Hungary, Iceland, Ireland, Italy, Japan, Luxembourg, Malta, Mongolia, Netherlands, New Zealand, Norway, Poland, Portugal, South Africa, Ukraine, USSR, United Kingdom, United States.

Abstaining.—Burma, China, Cuba, El Salvador, Greece, Indonesia, Iran, Israel, Ivory Coast, Laos, Lebanon, Liberia, Libya, Madagascar, Malawi, Nigeria, Philippines, Romania, Saudi Arabia, Sierra Leone, Spain, Sudan, Swaziland, Syria, Togo, Turkey, United Arab Republic, Upper Volta.

Absent.—Albania, Botswana, Cambodia, Cameroon, Equatorial Guinea, Gabon, Gambia, Senegal.

APPENDIX I

TREATY ON THE PROHIBITION OF THE EMPLACEMENT OF NUCLEAR WEAPONS AND OTHER WEAPONS OF MASS DESTRUCTION ON THE SEABED AND THE OCEAN FLOOR AND IN THE SUBSOIL THEREOF (1971)

The States Parties to this Treaty,

Recognizing the common interest of mankind in the progress of the exploration and use of the seabed and the ocean floor for peaceful purposes,

Considering that the prevention of a nuclear arms race on the seabed and the ocean floor serves the interests of maintaining world peace, reduces international tensions and strengthens friendly relations among States,

Convinced that this Treaty constitutes a step towards the exclusion of the seabed, the ocean floor and the subsoil thereof from the arms race,

Convinced that this Treaty constitutes a step towards a treaty on general and complete disarmament under strict and effective international control, and determined to continue negotiations to this end,

Convinced that this Treaty will further the purposes and principles of the Charter of the United Nations, in a manner consistent with the principles of international law and without infringing the freedoms of the high seas.

Have agreed as follows:

Article 1

1. The States Parties to this Treaty undertake not to emplant or emplace on the seabed and the ocean floor and in the subsoil thereof beyond the outer limit of a seabed zone, as defined in article II, any nuclear weapons or any other types of weapons of mass destruction as well as structures, launching installations or any other facilities specifically designed for storing, testing or using such weapons.

2. The undertakings of paragraph 1 of this article shall also apply to the seabed zone referred to in the same paragraph, except that within such seabed zone, they shall not apply either to the coastal State or to the seabed beneath its territorial waters.

3. The States Parties to this Treaty undertake not to assist, encourage or induce any State to carry out activities referred to in paragraph 1 of this article and not to participate in any other way in such actions.

Article II

For the purpose of this Treaty, the outer limit of the seabed zone referred to in article I shall be coterminous with the twelve-mile outer limit of the zone referred to in part II of the Convention on the Territorial Sea and the Contiguous Zone, signed at Geneva on April 29, 1958, and shall be measured in accordance with the provisions of part I, section II, of that Convention and in accordance with international law.

Article III

1. In order to promote the objectives of and insure compliance with the provisions of this Treaty, each State Party to the Treaty shall have the right to verify through observation the activities of other States Parties to the Treaty on the seabed and the ocean floor and in the subsoil thereof beyond the zone referred to in article I, provided that observation does not interfere with such activities.

2. If after such observation reasonable doubts remain concerning the fulfillment of the obligations assumed under the Treaty, the State Party having such doubts and the State Party that is responsible for the activities giving rise to the doubts shall consult with a view to removing the doubts. If the doubts persist, the State Party having such doubts shall notify the other States Parties, and the Parties concerned shall cooperate on such further procedures for verification as may be agreed, including appropriate inspection of objects, structures, installations or other facilities that reasonably may be expected to be of a kind described in article I. The Parties in the region of the activities, including any coastal State, and any other Party so requesting, shall be entitled to participate in such consultation and cooperation. After completion of the further procedures for verification, an appropriate report shall be circulated to other Parties by the Party that initiated such procedures.

3. If the State responsible for the activities giving rise to the reasonable doubts is not identifiable by observation to the object, structure, installation or other facility, the State Party having such doubts shall notify and make appropriate inquiries of States Parties in the region of the activities and of any other State Party. If it is

ascertained through these inquiries that a particular State Party is responsible for the activities, that State Party shall consult and cooperate with other Parties as provided in paragraph 2 of this article. If the identity of the State responsible for the activities cannot be ascertained through these inquiries, then further verification procedures, including inspection, may be undertaken by the inquiring State Party, which shall invite the participation of the Parties in the region of the activities, including any coastal State, and of any other Party desiring to cooperate.

4. If consultation and cooperation pursuant to paragraphs 2 and 3 of this article have not removed the doubts concerning the activities and there remains a serious question concerning fulfillment of the obligations assumed under this Treaty, a State Party may, in accordance with the provisions of the Charter of the United Nations, refer the matter to the Security Council, which may take action in accordance with the Charter.

5. Verification pursuant to this article may be undertaken by any State Party using its own means, or with the full or partial assistance of any other State Party, or through appropriate international procedures within the framework of the United Nations and in accordance with its Charter.

6. Verification activities pursuant to this Treaty shall not interfere with activities of other States Parties and shall be conducted with due regard for rights recognized under international law, including the freedoms of the high seas and the rights of coastal States with respect to the exploration and exploitation of their continental shelves.

Article IV

Nothing in this Treaty shall be interpreted as supporting or prejudicing the position of any State Party with respect to existing international conventions, including the 1958 Convention on the Territorial Sea and the Contiguous Zone, or with respect to rights or claims which such State Party may assert, or with respect to recognition or nonrecognition of rights or claims asserted by any other State, related to waters off its coasts, including, inter alia, territorial seas and contiguous zones, or to the seabed and the ocean floor, including continental shelves.

Article V

The Parties to this Treaty undertake to continue negotiations in good faith concerning further measures in the field of disarmament for the prevention of an arms race on the seabed, the ocean floor and the subsoil thereof.

Article VI

Any State Party may propose amendments to this Treaty. Amendments shall enter into force for each State Party accepting the amendments upon their acceptance by a majority of the States Parties to the Treaty and, thereafter, for each remaining State Party on the date of acceptance by it.

Article VII

Five years after the entry into force of this Treaty, a conference of Parties to the Treaty shall be held at Geneva, Switzerland, in order to review the operation of this Treaty with a view to assuring that the purposes of the preamble and the provisions of the Treaty are being realized. Such review shall take into account any relevant technological developments. The review conference shall determine, in accordance with the views of a majority of those Parties attending, whether and when an additional review conference shall be convened.

Article VIII

Each State Party to this Treaty shall in exercising its national sovereignty have the right to withdraw from this Treaty if it decides that extraordinary events related to the subject matter of this Treaty have jeopardized the supreme interests of its country. It shall give notice of such withdrawal to all other States Parties to the Treaty and to the United Nations Security Council three months in advance. Such notice shall include a statement of the extraordinary events it considers to have jeopardized its supreme interests.

Article IX

The provisions of this Treaty shall in no way affect the obligations assumed by States Parties to the Treaty under international instruments establishing zones free from nuclear weapons.

Article X

1. This Treaty shall be open for signature to all States. Any State which does not sign the Treaty before its entry into force in accordance with paragraph 3 of this article may accede to it at any time.

2. This Treaty shall be subject to ratification by signatory States. Instruments of ratification and of accession shall be deposited with the Governments of the United States of America, the United Kingdom of Great Britain and Northern Ireland, and the Union of Soviet Socialist Republics, which are hereby designated the Depositary Governments.

3. This Treaty shall enter into force after the deposit of instruments of ratification by twenty-two Governments, including the Governments designated as Depositary Governments of this Treaty.

4. For States whose instruments of ratification or accession are deposited after the entry into force of this Treaty, it shall enter into force on the date of the deposit of their instruments of ratification or accession.

5. The Depositary Governments shall promptly inform the Governments of all signatory and acceding States of the date of each signature, of the date of deposit of each instrument of ratification or of accession, of the date of the entry into force of this Treaty, and of the receipt of other notices.

6. This Treaty shall be registered by the Depositary Governments pursuant to Article 102 of the Charter of the United Nations.

Article XI

This Treaty, the English, Russian, French, Spanish and Chinese texts of which are equally authentic, shall be deposited in the archives of the Depositary Governments. Duly certified copies of this Treaty shall be transmitted by the Depositary Governments to the Governments of the States signatory and acceding thereto.

IN WITNESS WHEREOF the undersigned, being duly authorized thereto, have signed this Treaty.

DONE in triplicate, at the cities of Washington, London and Moscow, this eleventh day of February, one thousand nine hundred seventy-one.

APPENDIX J

PRESIDENT NIXON'S STATEMENT ON U.S. OCEANS POLICY (MAY 23, 1970)

The nations of the world are now facing decisions of momentous importance to man's use of the oceans for decades ahead. At issue is whether the oceans will be used rationally and equitably and for the benefit of mankind or whether they will become an arena of unrestrained exploitation and conflicting jurisdictional claims in which even the most advantaged states will be losers.

The issue arises now—and with urgency—because nations have grown increasingly conscious of the wealth to be exploited from the seabeds and throughout the waters above, and because they are also becoming apprehensive about ecological hazards of unregulated use of the oceans and seabeds. The stark fact is that the law of the sea is inadequate to meet the needs of modern technology and the concerns of the international community. If it is not modernized multilaterally, unilateral action and international conflict are inevitable.

This is the time, then for all nations to set about resolving the basic issues of the future regime for the oceans—and to resolve it in a way that redounds to the general benefit in the area of intensive exploitation that lies ahead. The United States as a major maritime power and a leader in ocean technology to unlock the riches of the ocean has a special responsibility to move this effort forward.

Therefore, I am today proposing that all nations adopt as soon as possible a treaty under which they would renounce all national claims over the natural resources of the seabed beyond the point where the high seas reach a depth of 200 meters (218.8 yards) and would agree to regard these resources as the common heritage of mankind.

The treaty should establish an international regime for the exploitation of seabed resources beyond this limit. The regime should provide for the collection of substantial mineral royalties to be used for international community purposes, particularly economic assistance to developing countries. It should also establish general rules to prevent unreasonable interference with other users of the ocean, to protect the ocean from pollution, to assure the integrity of the investment necessary for such exploitation, and to provide for peaceful and compulsory settlement of disputes.

I propose two types of machinery for authorizing exploitation of seabed resources beyond a depth of 200 meters.

First, I propose that coastal nations act as trustees for the international community in an international trusteeship zone comprised of the continental margins beyond a depth of 200 meters off their coasts. In return, each coastal state would receive a share of the international revenues from the zone in which it acts as trustee and could impose additional taxes if these were deemed desirable.

As a second step, agreed international machinery would authorize and regulate exploration and use of seabed resources beyond the continental margins.

The United States will introduce specific proposals at the next meeting of the United Nations Seabeds Committee to carry out these objectives.

Although I hope agreement on such steps can be reached quickly, the negotiation of such a complex treaty may take some time. I do not, however, believe it is either necessary or desirable to try to halt exploration and exploitation of the seabeds beyond a depth of 200 meters during the negotiating process.

Accordingly, I call on other nations to join the United States in an interim policy. I suggest that all permits for exploration and exploitation of the seabeds beyond 200 meters be issued subject to the international regime to be agreed upon. The regime should accordingly include due protection for the integrity of investments made in the interim period. A substantial portion of the revenues derived by a state from exploitation beyond 200 meters during this interim period should be turned over to an appropriate international development agency for assistance to developing countries. I would plan to seek appropriate Congressional action to make such funds available as soon as a sufficient number of other states also indicate their willingness to join this interim policy.

I will propose necessary changes in the domestic import and tax laws and regulations of the United States to assure that our own laws and regulations do not discriminate against U.S. nationals operating in the trusteeship zone off our coast or under the authority of the international machinery to be established.

It is equally important to assure unfettered and harmonious use of the oceans as an avenue of commerce and transportation, and as a source of food. For this reason the United States is currently engaged with other states in an effort to obtain a new law of the sea treaty. This treaty would establish a 12-mile limit for territorial seas and provide for free transit through international straits. It would also accommodate the problems of developing countries and other nations regarding the conservation and use of the living resources of the high seas.

I believe that these proposals are essential to the interests of all nations, rich and poor, coastal and landlocked, regardless of their political systems. If they result in international agreements, we can save over two-thirds of the earth's surface from national conflict and rivalry, protect it from pollution, and put it to use for the benefit of all. This would be a fitting achievement for this 25th anniversary year of the United Nations.

APPENDIX K

THE AUGUST 3, 1970 UNITED STATES DRAFT CONVENTION ON THE INTERNATIONAL SEABED AREA

Working Paper

The attached draft of a United Nations Convention on the International Seabed Area is submitted by the United States Government as a working paper for discussion purposes.

The draft Convention and its Appendices raise a number of questions with respect to which further detailed study is clearly necessary and do not necessarily represent the definitive views of the United States Government. The Appendices in particular are included solely by way of example.

UNITED NATIONS CONVENTION ON THE INTERNATIONAL SEABED AREA

CHAPTER I—BASIC PRINCIPLES

Article 1

1. The International Seabed Area shall be the common heritage of all mankind.

2. The International Seabed Area shall comprise all areas of the seabed and subsoil of the high seas[1] seaward of the 200 meter isobath adjacent to the coast of continents and islands.

3. Each Contracting Party shall permanently delineate the precise boundary of the International Seabed Area off its coast by straight lines not exceeding 60 nautical miles in length, following the general direction of the limit specified in paragraph 2. Such lines shall connect fixed points at the limit specified in paragraph 2, defined permanently by coordinates of latitude and longitude. Areas between or landward of such points may be deeper than 200 meters. Where a trench or trough deeper than 200 meters transects an area less than 200 meters in depth, a straight boundary line more than 60 nautical miles in length, but not exceeding the lesser of one fourth

of the length of that part of the trench or trough transecting the area 200 meters in depth or 120 nautical miles, may be drawn across the trench or trough.

4. Each Contracting Party shall submit the description of the boundary to the International Seabed Boundary Review Commission within five years of the entry into force of this Convention for such Contracting Party. Boundaries not accepted by the Commission and not resolved by negotiation between the Commission and the Contracting Party within one year shall be submitted by the Commission to the Tribunal in accordance with Section E of Chapter IV.

5. Nothing in this Article shall affect any agreement or prejudice the position of any Contracting Party with respect to the delimitation of boundaries between opposite or adjacent States in seabed landward of the International Seabed Area, or, with respect to any delimitation pursuant to Article 30.

Article 2

1. No State may claim or exercise sovereignty or sovereign rights over any part of the International Seabed Area or its resources. Each Contracting Party agrees not to recognize any such claim or exercise of sovereignty or sovereign rights.

2. No State has, nor may it acquire, any right, or interest in the International Seabed Area or its resources except as provided in this Convention.

Note:—The preceding Article is not intended to imply that States do not currently have rights under, or consistent with, the 1958 Geneva Convention on the Continental Shelf.

Article 3

The International Seabed Area shall be open to use by all States, without discrimination, except as otherwise provided in this Convention.

---

1. The United States has simulatenously proposed an international Convention which would, inter alia, fix the boundary between the territorial sea and the high seas at a maximum distance of 12 nautical miles from the coast.

Article 4

The International Seabed Area shall be reserved exclusively for peaceful purposes.

Article 5

1. The International Seabed Resource Authority shall use revenues it derives from the exploration and exploitation of the mineral resources of the International Seabed Area for the benefit of all mankind, particularly to promote the economic advancement of developing States Parties to this Convention, irrespective of their geographic location. Payments to the Authority shall be established at levels designed to ensure that they make a continuing and substantial contribution to such economic advancement, bearing in mind the need to encourage investment in exploration and exploitation and to foster efficient development of mineral resources.

2. A portion of these revenues shall be used, through or in cooperation with other international or regional organizations, to promote efficient, safe and economic exploitation of mineral resources of the seabed; to promote research on means to protect the marine environment; to advance other international efforts designed to promote safe and efficient use of the marine environment; to promote development of knowledge of the International Seabed Area; and to provide technical assistance to Contracting Parties or their nationals for these purposes, without discrimination.

Article 6

Neither this Convention nor any rights granted or exercised pursuant thereto shall affect the legal status of the superjacent waters as high seas, or that of the air space above those waters.

Article 7

All activities in the marine environment shall be conducted with reasonable regard for exploration and exploitation of the natural resources of the International Seabed Area.

Article 8

Exploration and exploitation of the natural resources of the International Seabed Area must not result in any unjustifiable interference with other activities in the marine environment.

Article 9

All activities in the International Seabed Area shall be conducted with strict and adequate safeguards for the protection of human life and safety and of the marine environment.

Article 10

All exploration and exploitation activities in the International Seabed Area shall be conducted by a Contracting Party or group of Contracting Parties or natural or juridical persons under its or their authority or sponsorship.

Article 11

1. Each Contracting Party shall take appropriate measures to ensure that those conducting activities under its authority or sponsorship comply with this Convention.
2. Each Contracting Party shall make it an offense for those conducting activities under its authority or sponsorship in the International Seabed Area to violate the provisions of this Convention. Such offenses shall be punishable in accordance with administrative or judicial procedures established by the Authorizing or Sponsoring Party.
3. Each Contracting Party shall be responsible for maintaining public order on manned installations and equipment operated by those authorized or sponsored by it.
4. Each Contracting Party shall be responsible for damages caused by activities which it authorizes or sponsors to any other Contracting Party or its nationals.
5. A group of States acting together, pursuant to agreement among them or through an international organization, shall be jointly and severally responsible under this Convention.

Article 12

All disputes arising out of the interpretation or application of this Convention shall be settled in accordance with provisions of Section E of Chapter IV.

CHAPTER II—GENERAL RULES

A. Mineral Resources

Article 13

1. All exploration and exploitation of the mineral deposits of the International Seabed Area shall be licensed by the International Seabed Resource Authority or the appropriate Trustee Party. All licenses shall be subject to the provisions of this Convention.

2. Detailed rules to implement this Chapter are contained in Appendices A, B and C.

Article 14

1. There shall be fees for licenses for mineral exploration and exploitation.

2. The fees referred to in paragraph 1 shall be reasonable and be designed to defray the administrative expenses of the International Seabed Resource Authority and of the Contracting Parties in discharging their responsibilities in the International Seabed Area.

Article 15

1. An exploitation license shall specify the minerals or categories of minerals and the precise area to which it applies. The categories established shall be those which will best promote simultaneous and efficient exploitation of different minerals.

2. Two or more licensees to whom licenses have been issued for different materials in the same or overlapping areas shall not unjustifiably interfere with each other's activities.

Article 16

The size of the area to which an exploitation license shall apply and the duration of the license shall not exceed the limits provided for in this Convention.

Article 17

Licensees must meet work requirements specified in this Convention as a condition of retaining an exploitation license prior to and after commercial production is achieved.

Article 18

Licensees shall submit work plans and production plans, as well as reports and technical data acquired under an exploitation license, to the Trustee Party or the Sponsoring Party, as appropriate, and, to the extent specified by this Convention, to the International Seabed Resource Authority.

Article 19

1. Each Contracting Party shall be responsible for inspecting, at regular intervals, the activities of licensees authorized or sponsored by it. Inspection reports shall be submitted to the International Seabed Resource Authority.

2. International Seabed Resource Authority, on its own initiative or at the request of any interested Contracting Party, may inspect any licensed activity in cooperation with the Trustee Party or Sponsoring Party, as appropriate, in order to ascertain that the licensed operation is being conducted in accordance with this Convention. In the event the International Seabed Resource Authority believes that a violation of this Convention has occurred, it shall inform the Trustee Party or Sponsoring Party, as appropriate, and request that suitable action be taken. If, after a reasonable period of time, the alleged violation continues, the International Seabed Resource Authority may bring the matter before the Tribunal in accordance with Section E of Chapter IV.

Article 20

1. Licenses issued pursuant to this Convention may be revoked only for cause in accordance with the provisions of this Convention.

2. Expropriation of investments made, or unjustifiable interference with operations conducted, pursuant to a license is prohibited.

Article 21

1. Due notice must be given, by Notices to Mariners or other recognized means of notification, of the construction or deployment

of any installations or devices for the exploration or exploitation of mineral deposits, and permanent means for giving warning of their presence must be maintained. Any installations or devices extending into the superjacent waters which are abandoned or disused must be entirely removed.

2. Such installations and devices shall not possess the status of islands and shall have no territorial sea of their own.

3. Installations or devices may not be established where interference with the use of recognized sea lanes or airways is likely to occur.

B. Living Resources of the Seabed

Article 22

Subject to the provisions of Chapter III, each Contracting Party may explore and exploit the seabed living resources of the International Seabed Area in accordance with such conservation measures as are necessary to protect the living resources of the International Seabed Area and to maximize their growth and utilization.

C. Protection of the Marine Environment, Life and Property

Article 23

1. In the International Seabed Area, the International Seabed Resource Authority shall prescribe Rules and Recommended Practices, in accordance with Chapter V of this Convention, to ensure:

   a. The protection of the marine environment against pollution from exploration and exploitation activities such as drilling, dredging, excavation, disposal of waste, construction and operation or maintenance of installations and pipelines and other devices;
   b. The prevention of injury to persons, property and marine resources arising from the aforementioned activities;
   c. The prevention of any unjustifiable interference with other activities in the marine environment arising from the aforementioned activities.

2. Deep drilling in the International Seabed Area shall be undertaken only in accordance with the provisions of this Convention.

D. Scientific Research

Article 24

1. Each Contracting Party agrees to encourage, and to obviate interference with, scientific research.

2. The Contracting Parties shall promote international cooperation in scientific research concerning the International Seabed Area:

a. By participating in international programs and by encouraging cooperation in scientific research by personnel of different countries;

b. Through effective publication of research programs and the results of research through international channels;

c. By cooperation in measures to strengthen the research capabilities of developing countries, including the participation of their nationals in research programs.

E. International Marine Parks and Preserves

Article 25

In consultation with the appropriate international organizations or agencies, the International Seabed Resource Authority may designate as international marine parks and preserves specific portions of the International Seabed Area that have unusual educational, scientific or recreational value. The establishment of such a park or preserve in the International Trusteeship Area shall require the approval of the appropriate Trustee Party.

CHAPTER III—THE INTERNATIONAL TRUSTEESHIP

Article 26

1. The International Trusteeship Area is that part of the International Seabed Area comprising the continental or island margin between the boundary described in Article I and a line, beyond the base of the continental slope, or beyond the base of the slope of an island situated beyond the continental slope, where the downward inclination of the surface of the seabed declines to a gradient of 1:———.

2. Each Trustee Party shall permanently delineate the precise seaward boundary of the International Trusteeship Area off its coast by straight lines not exceeding 60 nautical miles in length, following the general direction of the limits specified in paragraph 1. Such

lines shall connect fixed points at the limit specified in paragraph 1, defined permanently by coordinates of latitude and longitude. Areas between or landward of such points may have a surface gradient of less than 1:———. Where an elongate basin or plain having a surface gradient of less than 1:——— transects an area having a gradient of more than 1:———, a straight boundary line more than 60 nautical miles in length, but not exceeding the lesser of one-fourth of the length of that part of the basin or plain transecting the area having a gradient of more than 1:——— or 120 nautical miles, may be drawn across the basin or plain.[2]

3. Each Trustee Party shall submit the description of its boundary to the International Seabed Boundary Review Commission within five years of the entry into force of this Convention for that Party. Boundaries not accepted by that Commission and not resolved by negotiation between the Commission and the Trustee Party within one year shall be submitted by the Commission to the Tribunal for adjudication in accordance with Section E of Chapter IV.

(Note—Additional consideration will be given to problems raised by enclosed and semienclosed seas.)

Article 27

1. Except as specifically provided for in this Chapter, the coastal State shall have no greater rights in the International Trusteeship Area off its coast than any other Contracting Party.

2. With respect to exploration and exploitation of the natural resources of that part of the International Trusteeship Area in which it acts as trustee for the international community, each coastal State, subject to the provisions of this Convention, shall be responsible for:

a. Issuing, suspending and revoking mineral exploration and exploitation licenses;

b. Establishing work requirements, provided that such requirements shall not be less than those specified in Appendix A;

c. Ensuring that its licensees comply with this Convention, and, if it deems it necessary, applying standards to its

---

2. The precise gradient should be determined by technical experts, taking into account, among other factors, ease of determination, the need to avoid dual administration of single mineral deposits, and the avoidance of including excessively large areas in the International Trusteeship Area.

licensees higher than or in addition to those required under this Convention, provided such standards are promptly communicated to the International Seabed Resource Authority;

d. Supervising its licensees and their activities;

e. Exercising civil and criminal jurisdiction over its licensees, and persons acting on their behalf, while engaged in exploration or exploitation;

f. Filing reports with the International Seabed Resource Authority;

g. Collecting and transferring to the International Seabed Resource Authority all payments required by this Convention;

h. Determining the allowable catch of the living resources of the seabed and prescribing other conservation measures regarding them;

i. Enacting such laws and regulations as are necessary to perform the above functions.

3. Detailed rules to implement this Chapter are contained in Appendix C.

Article 28

In performing the functions referred to in Article 27, the Trustee Party may, in its discretion:

a. Establish the procedures for issuing licenses;

b. Decide whether a license shall be issued;

c. Decide to whom a license shall be issued, without regard to the provisions of Article 3;

d. Retain [a figure between 33 1/3% and 50% will be inserted here] of all fees and payments required by this Convention;

e. Collect and retain additional license and rental fees to defray its administrative expenses, and collect, and retain [a figure between 33 1/3% and 50% will be inserted here] of, other additional fees and payments related to the issuance or retention of a license, with annual notification to the International Seabed Resource Authority of the total amount collected;

f. Decide whether and by whom the living resources of the seabed shall be exploited, without regard to the provisions of Article 3.

Article 29

The Trustee Party may enter into an agreement with the International Seabed Resource Authority under which the International Seabed Resource Authority will perform some or all of the trusteeship supervisory and administrative functions provided for in this Chapter in return for an appropriate part of the Trustee Party's share of international fees and royalties.

Article 30

Where a part of the International Trusteeship Area is off the coast of two or more Contracting Parties, such Parties shall, by agreement, precisely delimit the boundary separating the areas in which they shall respectively perform their trusteeship functions and inform the International Seabed Boundary Review Commission of such delimitation. If agreement is not reached within three years after negotiations have commenced, the International Seabed Boundary Review Commission shall be requested to make recommendations to the Contracting Parties concerned regarding such delimitation. If agreement is not reached within one year after such recommendations are made, the delimitation recommended by the Commission shall take effect unless either Party, within 90 days thereafter, brings the matter before the Tribunal in accordance with Section E of Chapter IV.

CHAPTER IV—THE INTERNATIONAL SEABED RESOURCE AUTHORITY

A. General

Article 31

1. The International Seabed Resource Authority is hereby established.

2. The principal organs of the Authority shall be the Assembly, the Council, and the Tribunal.

Article 32

The permanent seat of the Authority shall be at ——.

Article 33

Each Contracting Party shall recognize the juridical personality of the Authority. The legal capacity, privileges and immunities of the Authority shall be the same as those defined in the Convention on the Privileges and Immunities of the Specialized Agencies of the United Nations.

B. The Assembly

Article 34

1. The Assembly shall be composed of all Contracting Parties.
2. The first session of the Assembly shall be convened———. The Assembly shall thereafter be convened by the Council at least once every three years at a suitable time and place. Extraordinary sessions of the Assembly shall be convened at any time on the call of the Council, or the Secretary-General of the Authority at the request of one-fifth of the Contracting Parties.
3. At meetings of the Assembly a majority of the Contracting Parties is required to constitute a quorum.
4. In the Assembly each Contracting Party shall exercise one vote.
5. Decisions of the Assembly shall be taken by a majority of the members present and voting, except as otherwise provided in this Convention.

Article 35

The powers and duties of the Assembly shall be to:

a. Elect its President and other officers;

b. Elect members of the Council in accordance with Article 36;

c. Determine its rules of procedure and constitute such subsidiary organs as it considers necessary or desirable;

d. Require the submission of reports from the Council;

e. Take action on any matter referred to it by the Council;

f. Approve proposed budgets for the Authority, or return them to the Council for reconsideration and resubmission;

g. Approve proposals by the Council for changes in the allocation of the net income of the Authority within the limits prescribed in Appendix D, or return them to the Council for reconsideration and resubmission;

h. Consider any matter within the scope of this Convention and make recommendations to the Council or Contracting Parties as appropriate;

i. Delegate such of its powers as it deems necessary or desirable to the Council and revoke or modify such delegation at any time;

j. Consider proposals for amendments of this Convention in accordance with Article 76.

C. The Council

Article 36

1. The Council shall be composed of twenty-four Contracting Parties and shall meet as often as necessary.

2. Members of the Council shall be designated or elected in the following categories:

   a. The six most industrially advanced Contracting Parties shall be designated in accordance with Appendix E;

   b. Eighteen additional Contracting Parties, of which at least twelve shall be developing countries, shall be elected by the Assembly, taking into account the need for equitable geographical distribution.

3. At least two of the twenty-four members of the Council shall be landlocked or shelf-locked countries.

4. Elected members of the Council shall hold office for three years following the last day of the Assembly at which they are elected and thereafter until their successors are elected. Designated members of the Council shall hold office until replaced in accordance with Appendix E.

5. Representatives on the Council shall not be employees of the Authority.

Article 37

1. The Council shall elect its President for a term of three years.

2. The President of the Council may be a national of any Contracting Party, but may not serve during his term of office as its representative in the Assembly or on the Council.

3. The President shall have no vote.

4. The President shall:

   a. Convene and conduct meetings of the Council;

   b. Carry out the functions assigned to him by the Council.

Article 38

Decisions by the Council shall require approval by a majority of all its members, including a majority of members in each of the two categories referred to in paragraph 2 of Article 36.

Article 39

Any Contracting Party not represented on the Council may participate, without a vote, in the consideration by the Council or any of the subsidiary organs, of any question which is of particular interest to it.

Article 40

The powers and duties of the Council shall be to:

a. Submit annual reports to the Contracting Parties;
b. Carry out the duties specified in this Convention and any duties delegated to it by the Assembly;
c. Determine its rules of procedure;
d. Appoint and supervise the Commissions provided for in this Chapter, establish procedures for the coordination of their activities, and determine the terms of office of their members;
e. Establish other subsidiary organs, as may be necessary or desirable, and define their duties;
f. Appoint the Secretary-General of the Authority and establish general guidelines for the appointment of such other personnel as may be necessary;
g. Submit proposed budgets to the Assembly for its approval, and supervise their execution;
h. Submit proposals to the Assembly for changes in the allocation of the net income of the Authority within the limits prescribed in Appendix D;
i. Adopt and amend Rules and Recommended Practices in accordance with Chapter V, upon the recommendation of the Rules and Recommended Practices Commission;
j. Issue emergency orders, at the request of any Contracting Party, to prevent serious harm to the marine environment arising out of any exploration or exploitation activity and communicate them immediately to licensees, and Authorizing or Sponsoring Parties, as appropriate;
k. Establish a fund to provide emergency relief and assistance in the event of a disaster to the marine environment resulting from exploration or exploitation activities;

l. Establish procedures for coordination between the International Seabed Resource Authority, and the United Nations, its specialized agencies and other international or regional organizations concerned with the marine environment;

m. Establish or support such international or regional centers, through or in cooperation with other international and regional organizations, as may be appropriate to promote study and research of the natural resources of the seabed and to train nationals of any Contracting Party in related science and the technology of seabed exploration and exploitation, taking into account the special needs of developing States Parties to this Convention;

n. Authorize and approve agreements with a Trustee Party, pursuant to Article 29, under which the International Seabed Resource Authority will perform some or all of the Trustee Party's functions.

Article 41

In furtherance of Article 5, paragraph 2, of this Convention, the Council may, at the request of any Contracting Party and taking into account the special needs of developing States Parties to this Convention:

a. Provide technical assistance to any Contracting Party to further the objectives of this Convention;

b. Provide technical assistance to any Contracting Party to help it to meet its responsibilities and obligations under this Convention;

c. Assist any Contracting Party to augment its capability to derive maximum benefit from the efficient administration of the International Trusteeship Area.

D. The Commissions

Article 42

1. There shall be a Rules and Recommended Practices Commission, an Operations Commission, and an International Seabed Boundary Review Commission.

2. Each Commission shall be composed of five to nine members appointed by the Council from among persons nominated by Contracting Parties. The Council shall invite all Contracting Parties to submit nominations.

3. No two members of a Commission may be nationals of the same State.

4. A member of each Commission shall be elected its President by a majority of the members of the Commission.

5. Each Commission shall perform the functions specified in this Convention and such other functions as the Council may specify from time to time.

Article 43

1. Members of the Rules and Recommended Practices Commission shall have suitable qualifications and experience in seabed resources management, ocean sciences, maritime safety, ocean and marine engineering, and mining and mineral technology and practices. They shall not be full-time employees of the Authority.

2. The Rules and Recommended Practices Commission shall:
   a. Consider, and recommend to the Council for adoption, Annexes to this Convention in accordance with Chapter V;
   b. Collect from and communicate to Contracting Parties information which the Commission considers necessary and useful in carrying out is functions.

Article 44

1. Members of the Operations Commission shall have suitable qualifications and experience in the management of seabed resources, and operation of marine installations, equipment and devices.

2. The Operations Commission shall:
   a. Issue licenses for seabed mineral exploration and exploitation, except in the International Trusteeship Area;
   b. Supervise the operations of licensees in cooperation with the Trustee or Sponsoring Party, as appropriate, but shall not itself engage in exploration or exploitation;
   c. Perform such functions with respect to disputes between Contracting Parties as are specified in Section E of this Chapter;
   d. Initiate proceedings pursuant to Section E of this Chapter for alleged violations of this Convention, including but not limited to proceedings for revocation or suspension of licenses;
   e. Arrange for and review the collection of international fees and other forms of payment;
   f. Arrange for the collection and dissemination of information relating to licensed operations;
   g. Supervise the performance of the functions of the Authority pursuant to any agreement between a Trustee Party and the Authority under Article 29;

h. Issue deep drilling permits.

Article 45

1. Members of the International Seabed Boundary Review Commission shall have suitable qualifications and experience in marine hydrography, bathymetry, geodesy and geology. They shall not be full-time employees of the Authority.

2. The International Seabed Boundary Review Commission shall:

a. Review the delineation of boundaries submitted by Contracting Parties in accordance with Articles 1 and 26 to see that they conform to the provisions of this Convention, negotiate any differences with Contracting Parties, and if these differences are not resolved initiate proceedings before the Tribunal in accordance with Section E of this Chapter;

b. Make recommendations to the Contracting Parties in accordance with Article 30;

c. At the request of any Contracting Party, render advice on any boundary question arising under this Convention.

E. The Tribunal

Article 46

1. The Tribunal shall decide all disputes and advise on all questions relating to the interpretation and application of this Convention which have been submitted to it in accordance with the provisions of this Convention. In its decisions and advisory opinions the Tribunal shall also apply relevant principles of international law.

2. Subject to an authorization under Article 96 of the Charter of the United Nations, the Tribunal may request the International Court of Justice to give an advisory opinion on any question of international law.

Article 47

1. The Tribunal shall be composed of five, seven, or nine independent judges, who shall possess the qualifications required in their respective countries for appointment to the highest judicial offices, or shall be lawyers especially competent in matters within the scope of this Convention. In the Tribunal as a whole the representation of the principal legal systems of the world shall be assured.

2. No two of the members of the Tribunal may be nationals of the same State.

Article 48

1. Each Contracting Party shall be entitled to nominate candidates for membership on the Tribunal. The Council shall elect the Tribunal from a list of these nominations.

2. The members of the Tribunal shall be elected for nine years and may be re-elected, providing, however, that the Council may establish procedures for staggered terms. Should such procedures be established, the judges whose terms are to expire in less than nine years shall be chosen by lots drawn by the Secretary-General.

3. The members of the Tribunal shall continue to discharge their duties until their places have been filled. Though replaced, they shall finish any cases which they may have begun.

4. A member of the Tribunal unable to perform his duties may be dismissed by the Council on the unanimous recommendation of the other members of the Tribunal.

5. In case of a vacancy, the Council shall elect a successor who shall hold office for the remainder of his predecessor's term.

Article 49

The Tribunal shall establish its rules of procedure; elect its President; appoint its Registrar and determine his duties and terms of service; and adopt regulations for the appointment of the remainder of its staff.

Article 50

1. Any Contracting Party which considers that another Contracting Party has failed to fulfill any of its obligations under this Convention may bring its complaint before the Tribunal.

2. Before a Contracting Party institutes such proceedings before the Tribunal it shall bring the matter before the Operations Commission.

3. The Operations Commission shall deliver a reasoned opinion in writing after the Contracting Parties concerned have been given the opportunity both to submit their own cases and to reply to each other's case.

4. If the Contracting Party accused of a violation does not comply with the terms of such opinion within the period laid down by

the Commission, the other Party concerned may bring the matter before the Tribunal.

5. If the Commission has not given an opinion within a period of three months from the date when the matter was brought before it, either Party concerned may bring the matter before the Tribunal without waiting further for the opinion of the Commission.

Article 51

1. Whenever the Operations Commission, acting on its own initiative or at the request of any licensee, considers that a Contracting Party or a licensee has failed to fulfill any of its obligations under this Convention, it shall issue a reasoned opinion in writing on the matter after giving such party the opportunity to submit its comments.

2. If the party concerned does not comply with the terms of such opinion within the period laid down by the Commission, the latter may bring a complaint before the Tribunal.

Article 52

1. If the Tribunal finds that a Contracting Party or a licensee has failed to fulfill any of its obligations under this Convention, such party shall take the measures required for the implementation of the judgment of the Tribunal.

2. When appropriate, the Tribunal may decide that the Contracting Party or the licensee who has failed to fulfill its obligations under this Convention shall pay to the Authority a fine of not more than $1,000 for each day of the offense, or shall pay damages to the other party concerned, or both.

3. In the event the Tribunal determines that a licensee has committed a gross and persistent violation of the provisions of this Convention and has not within a reasonable time brought his operations into compliance with them, the Council may, as appropriate, either revoke his license or request that the Trustee Party revoke it. The licensee shall not, however, be deprived of his license if his actions were directed by a Trustee or Sponsoring Party.

Article 53

If disputes under Articles 1, 26 and 30 have not been resolved by the time and methods specified in those Articles, the International Seabed Boundary Review Commission shall bring the matter before the Tribunal.

Article 54

1. Any Contracting Party, which questions the legality of measures taken by the Council, the Rules and Recommended Practices Commission, the Operations Commission, or the International Seabed Boundary Review Commission on the grounds of a violation of this Convention, lack of jurisdiction, infringement of important procedural rules, unreasonableness, or misuse of powers, may bring the matter before the Tribunal.

2. Any person may, subject to the same conditions, bring a complaint to the Tribunal with regard to a decision directed to that person, or a decision which, although in the form of a rule or a decision directed to another person, is of direct concern to the Complainant.

3. The proceedings provided for in this Article shall be instituted within a period of two months, dating, as the case may be, either from the publication of the measure concerned or from its notification to the complainant, or, in default thereof, from the day on which the latter learned of it.

4. If the Tribunal considers the appeal well-founded, it should declare the measure concerned to be null and void, and shall decide to what extent the annulment shall have retroactive application.

Article 55

1. The organ responsible for a measure declared null and void by the Tribunal shall be required to take the necessary steps to comply with the Tribunal's judgment.

2. When appropriate, the Tribunal may require that the Authority repair or pay for any damage caused by its organs or by its officials in the performance of their duties.

Article 56

When a case pending before a court of tribunal of one of the Contracting Parties raises a question of the interpretation of this Convention or of the validity or interpretation of measures taken by an organ of the Authority, the court or tribunal concerned may request the Tribunal to give its advice thereon.

Article 57

The Tribunal shall also be competent to decide any dispute connected with the subject matter of this Convention submitted to it pursuant to an agreement, license, or contract.

Article 58

If a Contracting Party fails to perform the obligations incumbent upon it under a judgment rendered by the Tribunal, the other Party to the case may have recourse to the Council, which shall decide upon measures to be taken to give effect to the judgment. When appropriate, the Council may decide to suspend temporarily, in whole or in part, the rights under this Convention of the Party failing to perform its obligations, without impairing the rights of licensees who have not contributed to the failure to perform such obligations. The extent of such a suspension should be related to the extent and seriousness of the violation.

Article 59

In any case in which the Council issues an order in emergency circumstances to prevent serious harm to the marine environment, any directly affected Contracting Party may request immediate review by the Tribunal, which shall promptly either confirm or suspend the application of the emergency order pending the decision of the case.

Article 60

Any organ of the International Seabed Resource Authority may request the Tribunal to give an advisory opinion on any legal question connected with the subject matter of this Convention.

F. The Secretariat

Article 61

The Secretariat shall comprise a Secretary-General and such staff as the International Seabed Resource Authority may require. The Secretary-General shall be appointed by the Council from among persons nominated by Contracting Parties. He shall serve for a term of six years, and may be reappointed.

Article 62

The Secretary-General shall:

a. Be the chief administrative officer of the International Seabed Resource Authority, and act in that capacity in all meetings of the Assembly and the Council;

b. Report to the Assembly and the Council on the work of the International Seabed Resource Authority;

c. Collect, publish and disseminate information which will contribute to mankind's knowledge of the seabed and its resources;

d. Perform such other functions as are entrusted to him by the Assembly or the Council.

Article 63

1. In the performance of their duties the Secretary-General and the staff shall not seek or receive instructions from any government or from any other external authority. They shall refrain from any action which might reflect on their position as international officials responsible only to the International Seabed Resource Authority.

2. Each Contracting Party shall respect the exclusively international character of the responsibilities of the Secretary-General and the staff and shall not seek to influence them in the discharge of their responsibilities.

Article 64

1. The staff of the International Seabed Resource Authority shall be appointed by the Secretary-General under the general guidelines established by the Council.

2. Appropriate staffs shall be assigned to the various organs of the Authority as required.

3. The paramount consideration in the employment of the staff and in the determination of the conditions of service shall be the necessity of securing the highest standards of efficiency, competence, and integrity. Due regard shall be paid to the importance of recruiting the staff on as wide a geographical basis as possible.

G. Conflicts of Interest

Article 65

No representative to the Assembly or the Council nor any member of the Tribunal, Commissions, subsidiary organs (other than advisory bodies or consultants), or the Secretariat, shall, while serving as such a representative or member, be actively associated with or financially interested in any of the operations of any enterprise concerned with exploration or exploitation of the natural resources of the International Seabed Area.

CHAPTER V—RULES AND RECOMMENDED PRACTICES

Article 66

1. Rules and Recommended Practices are contained in Annexes to this Convention.

2. Annexes shall be consistent with this Convention, its Appendices, and any amendments thereto. Any Contracting Party may challenge an Annex, an amendment to an Annex, or any of their provisions, on the grounds that it is unnecessary, unreasonable or constitutes a misuse of powers, by bringing the matter before the Tribunal in accordance with Article 54.

3. Annexes shall be adopted and amended in accordance with Article 67. Those annexes adopted along with this Convention, if any, may be amended in accordance with Article 67.

Article 67

The Annexes to this Convention and amendments to such Annexes shall be adopted in accordance with the following procedure:

a. They shall be prepared by the Rules and Recommended Practices Commission and submitted to the Contracting Parties for comments;

b. After receiving the comments, the Commission shall prepare a revised text of the Annex or amendments thereto;

c. The text shall then be submitted to the Council which shall adopt it or return it to the Commission for further study;

d. If the Council adopts the text, it shall submit it to the Contracting Parties;

e. The Annex or an amendment thereto shall become effective within three months after its submission to the Contracting

Parties, or at the end of such longer period of time as the Council may prescribe, unless in the meantime more than one-third of the Contracting Parties register their disapproval with the Authority;

f. The Secretary-General shall immediately notify all Contracting States of the coming into force of any Annex or amendment thereto.

Article 68

1. Annexes shall be limited to the Rules and Recommended Practices necessary to:

a. Fix the level, basis, and accounting procedures for determining international fees and other forms of payment, within the ranges specified in Appendix A;

b. Establish work requirements within the ranges specified in Appendices A and B;

c. Establish criteria for defining the technical and financial competence of applicants for licenses;

d. Assure that all exploration and exploitation activities, and all deep drilling, are conducted with strict and adequate safeguards for the protection of human life and safety and of the marine environment;

e. Protect living marine organisms from damage arising from exploration and exploitation activities;

f. Prevent or reduce to acceptable limits interference arising from exploration and exploitation activities with other uses and users of the marine environment;

g. Assure safe design and construction of fixed exploration and exploitation installations and equipment;

h. Facilitate search and rescue services, including assistance to aquanauts, and the reporting of accidents;

i. Prevent unnecessary waste in the extraction of minerals from the seabed;

j. Standardize the measurement of water depth and the definition of other natural features pertinent to the determination of the precise location of International Seabed Area boundaries;

k. Prescribe the form in which Contracting Parties shall describe their boundaries and the kinds of information to be submitted in support of them;

l. Encourage uniformity in seabed mapping and charting;

m. Facilitate the management of a part of the international trusteeship area pursuant to any agreement between a Trustee Party and the Authority under Article 29;

n. Establish and prescribe conditions for the use of international marine parks and preserves;

2. Application of any Rule or Recommended Practice may be limited as to duration or geographic area, but without discrimination against any Contracting Party or licensee.

Article 69

The Contracting Parties agree to collaborate with each other and the appropriate Commission in securing the highest practicable degree of uniformity in regulations, standards, procedures and organizations in relation to the matters covered by Article 68 in order to facilitate and improve seabed resources exploration and exploitation.

Article 70

Annexes and amendments thereto shall take into account existing international agreements and, where appropriate, shall be prepared in collaboration with other competent international organizations. In particular, existing international agreements and regulations relating to safety of life at sea shall be respected.

Article 71

1. Except as otherwise provided in this Convention, the Annexes and amendments thereto adopted by the Council shall be binding on all Contracting Parties.
2. Recommended Practices shall have no binding effect.

Article 72

Any Contracting Party believing that a provision of an Annex or an amendment thereto cannot be reasonably applied because of special circumstances may seek a waiver from the Operations Commission and if such waiver is not granted within three months, it may appeal to the Tribunal within an additional period of two months.

CHAPTER VI—TRANSITION

Article 73

1. There shall be due protection for the integrity of investments made in the International Seabed Area prior to the coming into force of this Convention.

2. All authorizations by a Contracting Party to exploit the mineral resources of the International Seabed Area granted prior to July 1, 1970, shall be continued without change after the coming into force of this Convention provided that:

a. Activities pursuant to such authorizations shall, to the extent possible, be conducted in accordance with the provisions of this Convention;

b. New activities under such previous authorizations which are begun after the coming into force of this Convention shall be subject to the regulatory provisions of this Convention regarding the protection of human life and safety and of the marine environment and the avoidance of unjustifiable interference with other uses of the marine environment;

c. Upon the expiration or relinquishment of such authorizations, or upon their revocation by the authorizing Party, the provisions of this Convention shall become fully applicable to any exploration or exploitation of resources remaining in the areas included in such authorizations;

d. Contracting Parties shall pay to the International Seabed Resource Authority, with respect to such authorizations, the production payments provided for under this Convention.

3. A Contracting Party which has authorized exploitation of the mineral resources of the International Seabed Area on or after July 1, 1970, shall be bound, at the request of the person so authorized, either to issue new licenses under this Convention in its capacity as a Trustee Party, or to sponsor the application of the person so authorized to receive new licenses from the International Seabed Resource Authority. Such new license issued by a Trustee Party shall include the same terms and conditions as its previous authorization, provided that such license shall not be inconsistent with this Convention, and provided further that the Trustee Party shall itself be responsible for complying with increased obligations resulting from the application of this Convention, including fees and other payments required by this Convention.

4. The provisions of paragraph 3 shall apply within one year after this Convention enters into force for the Contracting Party

concerned, but in no event more than five years after the entry into force of this Convention.

5. Until converted into new licenses under paragraph 3, all authorizations issued on or after July 1, 1970, to exploit the mineral resources of the International Seabed Area shall have the same status as authorizations under paragraph 2. Five years after the entry into force of this Convention all such authorizations not converted into new licenses under paragraph 3 shall be null and void.

6. Any Contracting Party that has authorized activities within the International Seabed Area after July 1, 1970, but before this Convention has entered into force for such Party, shall compensate its licensees for any investment losses resulting from the application of this Convention.

Article 74

1. The membership of the Tribunal, the Commissions, and the Secretariat shall be maintained at a level commensurate with the tasks being performed.

2. In the period before the International Seabed Resource Authority acquires income sufficient for the payment of its administrative expenses, the Authority may borrow funds for the payment of those expenses. The Contracting Parties agree to give sympathetic consideration to requests by the Authority for such loans.

CHAPTER VII—DEFINITIONS

Article 75

Unless another meaning results from the context of a particular provision, the following definitions shall apply;

1. "Convention" refers to all provisions of and amendments to this Convention, its Appendices, and its Annexes.

2. "Trustee Party" refers to the Contracting Party exercising trusteeship functions in that part of the International Trusteeship Area off its coast in accordance with Chapter III.

3. "Sponsoring Party" refers to a Contracting Party which sponsors an application for a license or permit before the International Seabed Resource Authority. The term "sponsor" is used in this context.

4. "Authorizing Party" refers to a Contracting Party authorizing any activity in the International Seabed Area, including a Trustee Party

issuing exploration or exploitation licenses. The term "authorize" is used in this context. In the case of a vessel, the term "Authorizing Party" shall be deemed to refer to the State of its nationality.

5. "Operating Party" refers to a Contracting Party which itself explores or exploits the natural resources of the International Seabed Area.

6. "Licensee" refers to a State, group of States, or natural or juridical person holding a license for exploration or exploitation of the natural resources of the International Seabed Area.

7. "Exploration" refers to any operation in the International Seabed Area which has as its principal or ultimate purpose the discovery and appraisal, or exploitation, of mineral deposits, and does not refer to scientific research. The term does not refer to similar activities when undertaken pursuant to an exploitation license.

8. "Deep drilling" refers to any form of drilling or excavation in the International Seabed Area deeper than 300 meters below the surface of the seabed.

9. "Landlocked or shelf-locked country" refers to a Contracting Party which is not a Trustee Party.

CHAPTER VIII—AMENDMENT AND WITHDRAWAL

Article 76

Any proposed amendment to this Convention or the appendices thereto which has been approved by the Council and a two-thirds vote of the Assembly shall be submitted by the Secretary-General to the Contracting Parties for ratification in accordance with their respective constitutional processes. It shall come into force when ratified by two-thirds of the Contracting Parties, including each of the six States designated pursuant to subparagraph 2(a) of Article 36 at the time the Council approved the amendments. Amendments shall not apply retroactively.

Article 77

1. Any Contracting Party may withdraw from this Convention by a written notification addressed to the Secretary-General. The Secretary-General shall promptly inform the other Contracting Parties of any such withdrawal.

2. The withdrawal shall take effect one year from the date of the receipt by the Secretary-General of the notification.

APPENDIX A

Terms and Procedures Applying to All Licenses in the International Seabed Area

1. Activities Requiring a License or a Permit:

1.1. Pursuant to Article 13 of this Convention, all exploration and exploitation operations in the International Seabed Area which have as their principal or ultimate purpose the discovery or appraisal, and exploitation, or mineral deposits shall be licensed.

1.2. There shall be two categories of licenses:

(a) A non-exclusive exploration license shall authorize geophysical and geochemical measurements, and bottom sampling, for the purposes of exploration. This license shall not be restricted as to area and shall grant no exclusive right to exploration nor any preferential right in applying for an exploitation license. It shall be valid for two years following the date of its issuance and shall be renewable for successive two-year periods.

(b) An exploitation license shall authorize exploration and exploitation of one of the groups of minerals described in Section 5 of this Appendix in a specified area. The exploitation license shall include the exclusive right to undertake deep drilling and other forms of subsurface entry for the purpose of exploration and exploitation of minerals described in paragraphs 5.1(a) and 5.1(c). The license shall be for a limited period and shall expire at the end of fiteeen years if no commercial production is achieved.

1.3. The right to undertake deep drilling for exploration or exploitation shall be granted only under an exploitation license.

1.4. Deep drilling for purposes other than exploration or exploitation of seabed minerals shall be authorized under a deep-drilling permit issued at no charge by the International Seabed Resource Authority, provided that:

(a) The application is accompanied by a statement from the Sponsoring Party certifying as to the applicant's technical competence and accepting liability for any damages that may result from such drilling;

(b) The application for such a permit is accompanied by a description of the location proposed for such holes, by seismograms and other pertinent information on the geology in the vicinity of the proposed drilling sites, and by a description of the equipment and procedures to be utilized;

(c) The proposed drilling, including the methods and equipment to be utilized, complies with the requirements of this Convention and is judged by the Authority not to pose an uncontrollable hazard to human safety, property, and the environment;

(d) The proposed drilling is either not within an area already under an exploitation license or is not objected to by the holder of such a license;

(e) The applicant agrees to make available promptly the geologic information obtained from such drilling to the Authority and the public.

2. General License Procedures:

2.1. An Authorizing or Sponsoring Party shall certify the operator's financial and technical competence and shall require the operator to conform to the rules, provisions and procedures specified under the terms of the license.

2.2. Each Authorizing or Sponsoring Party shall formulate procedures to ensure that applications for licenses are handled expeditiously and fairly.

2.3. Any Authorizing or Sponsoring Party which considers that it is unable to exercise appropriate supervision over operators authorized or sponsored by it in accordance with this Convention shall be permitted to authorize or sponsor operators only if their operations are supervised by the International Seabed Resource Authority persuant to an agreement between the Authorizing or Sponsoring Party and the International Seabed Resource Authority. In such event fees and rentals normally payable to the International Seabed Resource Authority will be increased appropriately to offset its supervisory costs.

3. Exploration Licenses—Procedures:

3.1. All applications for exploration licenses and for their renewal shall be accompanied by a fee of from $500 to $1,500 as specified in an Annex and a description of the location of the general area to be investigated and the kinds of activities to be undertaken. A portion [a figure between 50% and 66 2/3% will be inserted here] of the fee shall be forwarded by the Authorizing or Sponsoring Party to the Authority together with a copy of the application.

3.2. The Authorizing or Sponsoring Party shall transmit to the Authority the description referred to in paragraph 3.1 and its assurance that the activities will not be harmful to the marine environment.

3.3. The Authorizing or Sponsoring Party may require the operator to pay, and may retain, an additional license fee not to exceed $3,000, to help cover the administrative expenses of that Party.

3.4. Exploration licenses shall not be renewed in the event the operator has failed to conform his activities under the prior license to the provisions of this Convention or to the conditions of the license.

4. Exploitation Licenses—Procedures:

4.1. All applications for exploitation licenses shall be accompanied by a fee of from $5,000 to $15,000, per block, as specified in an Annex. A portion [a figure between 50% and 66 2/3% will be inserted here] of the fee shall be forwarded by the Authorizing or Sponsoring Party to the Authority together with a copy of the application.

4.2. Pursuant to Section 5 of this Appendix, applications shall identify the category of minerals in the specific area for which a license is sought.

4.3. When a license is granted to an applicant for more than one block at the same time, only a single certificate need be issued.

4.4. The Authorizing or Sponsoring Party may require the operator to pay, and may retain, an additional license fee not to exceed $30,000, to help cover the administrative expenses of that Party.

4.5. The license fee described in paragraph 4.1 shall satisfy the first two years' rental fee.

5. Exploitation Rights—Categories and Size of Blocks:

5.1. Licenses to exploit shall be limited to one of the following categories of minerals:

(a) Fluids or minerals extracted in a fluid state, such as oil, gas, helium, nitrogen, carbon dioxide, water, geothermal energy, sulfur and saline minerals.

(b) Manganese-oxide nodules and other minerals at the surface of the seabed.

(c) Other minerals, including category (b) minerals that occur beneath the surface of the seabed and metalliferous muds.

5.2. An exploitation license shall be issued for a specific area of the seabed and subsoil vertically below it, hereinafter referred to as a "block". The methods for defining the boundaries of blocks, and of portions thereof, shall be specified in an Annex.

5.3. In the category described in paragraph 5.1(a) the block shall be approximately 500 square kilometers, which shall be reduced to a quarter of a block when production begins. Each exploitation license shall apply to not more than one block, but exploitation licenses to rectangle containing as many as 16 contiguous blocks may be taken out under a single certificate and reduced by three quarters to a number of blocks, a single block, or a portion of a single block when production begins. The relinquishment requirement shall not apply to licenses issued for areas of one quarter of a block or less.

5.4. In the category described in paragraph 5.1(b) the block shall be approximately 40,000 square kilometers, which shall be reduced to a quarter of a block when production begins. Each exploitation license shall apply to not more than one block, but exploitation licenses to a rectangle containing as many as four contiguous blocks may be taken out under a single certificate and reduced to a single block, or to a portion of a single block, comprising one-fourth their total area, when production begins. The relinquishment requirement shall not apply to licenses issued for areas of one quarter of a block or less.

5.5. In the category described in paragraph 5.1(c) the block shall be approximately 500 square kilometers, which shall be reduced to one eighth of a block when production begins. Each license shall apply to not more than one block, but exploitation licenses to as many as 8 contiguous blocks may be taken out under a single certificate and reduced to a single block, or to a portion of a single block, comprising one eighth their total area, when production begins. The relinquishment shall not apply to licenses issued for one eight of a block or less.

5.6. Applications for exploitation licenses may be for areas smaller than the maximum stated above.

5.7. Operators may at any time relinquish rights to all or part of the licensed area.

5.8. Commercial production shall be deemed to have commenced or to be maintained when the value at the site of minerals exploited is not less than $100,000 per annum. The required minimum and the method of ascertaining this value shall be determined by the Authority.

5.9. If the commercial production is not maintained, the exploitation license shall expire within five years of its cessation, but when production is interrupted or suspended for reasons beyond the operator's control, the duration of the license shall be extended by a time equal to the period in which production has been suspended for reasons beyond the operator's control.

6. Rental Fees and Work Requirements:

Rental Fees

6.1. Prior to attaining commercial production the following annual rental fees shall be paid beginning in the third year after the license has been issued: (a) $2-$10 per square kilometer, as specified in an appropriate Annex, for the category of minerals described in paragraph 5.1(a) above; (b) $2-$10 per 100 square kilometers for the category of minerals described in paragraph 5.1(b) above, and (c)

$2-$10 per square kilometer for the category of minerals described in paragraph 5.1(c) above.

6.2. The rates in paragraph 6.1 shall increase at the rate of 10% per annum, calculated on the original base rental fee, for the first ten years after the third year, and shall increase 20% per annum, calculated on the original base rental fee, for the following two years.

6.3. After commercial production begins, the annual rental fee shall be $5,000-$25,000 per block, regardless of block size.

6.4. The rental fee shall be payable annually in advance to the Authorizing or Sponsoring Party which shall forward a portion [a figure between 50% and 66 2/3% will be inserted here] of the fees to the Authority. The Authorizing or Sponsoring Party may require the operator to pay, and may retain, an additional rental fee, not to exceed an amount equal to the amount paid pursuant to paragraphs 6.1 through 6.3, to help cover the administration expenses of that Party.

Work Requirements

6.5. Prior to attaining commercial production, the operator shall deposit a work requirement fee, or post a sufficient bond for that amount, for each license at the beginning of each year.

6.6. The minimum annual work requirement fee for each block shall increase in accordance with the following schedule:

Paragraph 5.1 (a) and (c) minerals

Years:	Amounts per annum
1 to 5	$20,000
6 to 10	180,000
11 to 15	200,000
Total	2,000,000

Paragraph 5.1 (b) minerals

Years:	Amounts per annum
1 to 2	$20,000
3 to 10	120,000
11 to 15	200,000
Total	2,000,000

The minimum annual work requirement fee for a portion of a block shall be an appropriate fraction of the above, to be specified in an Annex.

6.7. The work requirement fee shall be refunded to the operator upon receipt of proof by the Authorizing Party or Sponsoring Party that the amount equivalent to the fee has been expended in actual operations. Expenditures for on-land design or process research and equipment purchase or off-site construction cost directly related to the licensed block or group of blocks shall be considered to apply toward work requirements up to 75% of the amount required.

6.8. Expenditures in excess of the required amount for any given year shall be credited to the requirement for the subsequent year or years.

6.9. In the absence of satisfactory proof that the required expenditure has been made in accordance with the foregoing provisions of this section, the deposit will be forfeited.

6.10. If cumulative work requirement expenditures are not met at the end of the initial five-year period, the exploitation license shall be forfeited.

6.11. After commercial production begins the operator shall make an annual deposit of at least $100,000 at the beginning of each year; or shall post a sufficient bond for that amount, which shall be refunded in an amount equivalent to expenditures on or related to the block and the value of production at the site.

6.12. If production is suspended or delayed for reasons beyond the operator's control, the operator shall not be required to make the deposit or post the bond required in paragraph 6.11.

7. Submission of Work Plans and Data Under Exploitation Licenses Prior to Commencement of Commercial Production:

7.1. Exploitation license applications shall be accompanied by a general description of the work to be done and the equipment and methods to be used. The license shall submit subsequent changes in his work plan to the Sponsoring or Authorizing Party for review.

7.2. The licensee shall furnish reports at specified intervals to the Authorizing or Sponsoring Party supplying proof that he has fulfilled the specified work requirements. Copies of such reports shall be forwarded to the Authority.

7.3. The licensee shall maintain records of drill logs, geophysical data and other data acquired in the area to which his license refers, and shall provide access to them to the Authorizing or Sponsoring Party on request.

7.4. At intervals of five years, or when he relinquishes his rights to all or part of the area or when he submits a production plan as described in Section 8 of this Appendix, the operator shall transmit to the Authorizing or Sponsoring Party such maps, seismic sections, logs, assays, or reports, as are specified in an Annex to this Convention. The Authorizing or Sponsoring Party shall hold such data in confidence for ten years after receipt, but shall make the data available on request to the Authority for its confidential use in the inspection of operations.

7.5. The data referred to in paragraph 7.4 shall be transmitted to the Authority ten years after receipt by the Authorizing or Sponsoring Party, and made available by the Authority for public inspection. Such data shall be transmitted to the Authority immediately upon revocation of a license.

8. Production Plan and Producing Operations:

8.1. Prior to beginning commercial production the licensee shall submit a production plan to the Authorizing or Sponsoring Party and through such Party to the Authority.

8.2. The Authorizing or Sponsoring Party and the Authority shall require such modifications in the plan as may be necessary for it to meet the requirements of this Convention.

8.3. Any change in the Licensee's plan shall be submitted to the Authorizing or Sponsoring Party and through such Party to the Authority for their review and approval.

8.4. Not later than three months after the end of each year from the issuance of the license the licensee shall transmit to the Authorizing or Sponsoring Party for forwarding to the Authority production reports and such other data as may be specified in an Annex to this Convention.

8.5. The operator shall maintain geologic, geophysical and engineering records and shall provide access to them to the Authorizing or Sponsoring Party on its request. In addition, the operator shall submit annually such maps, sections, and summary reports, as are specified in Annexes to this Convention.

8.6. The Sponsoring or Authorizing Party shall hold such maps and reports in confidence for ten years from the time received but shall make them available on request to the Authority for its confidential use in the inspection of operations.

8.7. Such maps and reports shall be transmitted to the Authority and shall be made available by it for public inspection not later than ten years after receipt by the Sponsoring or Authorizing Party.

9. Unit Operations:

9.1. Accumulations of fluids and other minerals that can be made to migrate from one block to another and that would be most rationally mined by an operation under the control of a single operator but that lie astride the boundary of adjacent blocks licensed to different operators shall be brought into unit management and production.

9.2. With respect to deposits lying astride the seaward boundary of the International Trusteeship Area, the Operations Commission shall assure unit management and production, giving the Trustee and Sponsoring Parties and their licensees a reasonable time to reach agreement on an operation plan.

10. Payments on Production:

10.1. When commercial production begins under an exploitation license, the operator shall pay a cash production bonus of $500,000 to $2,000,000 per block, as specified in an Annex to this Convention, to the Authorizing or Sponsoring Party.

10.2. Thereafter, the operator shall make payments to the Authorizing or Sponsoring Party which are proportional to production, in the nature of total payments ordinarily made to governments under similar conditions. Such payments shall be equivalent to 5 to 40 percent of the gross value at the site of oil and gas, and 2 to 20 percent of the gross value at the site of other minerals, as specified in an Annex to this Convention. The total annual payment shall not be less than the annual rental fee under paragraph 6.3.

10.3. The Sponsoring Party shall forward all payments under this section to the Authority. The Authorizing Party shall forward a portion [a figure between 50% and 66 2/3% will be inserted here] of such payments to the Authority.

11. Graduation of Payments According to Environment and Other Factors:

11.1. The levels of payments and work requirements, as well as the rates at which such payments and work requirements escalate over time, may be graduated to take account of probable risk and cost to the investor, including such factors as water depth, climate, volume of production, proximity to existing production, or other factors affecting the economic rent that can reasonably be anticipated from mineral production in a given area.

11.2. Any graduated levels and rates shall be described and categorized in an Annex in such a way as to affect all licenses in each category equally and not to discriminate against or favor individual Parties or groups of Parties, or their nationals.

11.3. Any increases in such levels of payments or requirements shall apply only to new licenses or renewals and not to those already in force.

12. Liability:

12.1. The operator and his Authorizing or Sponsoring Party, as appropriate, shall be liable for damage to other users of the marine environment and for clean-up and restoration costs of damage to the land environment.

12.2. The Authorizing or Sponsoring Party, as appropriate, shall require operators to subscribe to an insurance plan or provide other means of guaranteeing responsibility, adequate to cover the liability described in paragraph——.

(Note—More detailed provisions on liability should be included.)

13. Revocation:

13.1. In the event of revocation pursuant to Article 52 of this Convention, there shall be no reimbursement for any expense incurred by the licensee prior to the revocation. The licensee shall, however, have the right to recover installations or equipment within six months of the date of the revocation of his license. Any installations or devices not removed by that time shall be removed and disposed of by the Authority, or the Authorizing or Sponsoring Party, at the expense of the licensee.

14. International Fees and Payments:

14.1. The Authority shall specify the intervals at which fees and other payments collected by an Authorizing or Sponsoring Party shall be transmitted.

14.2. No Contracting Party shall impose or collect any tax, direct or indirect, on fees and other payments to the Authority.

14.3. All fees and payments required under this Convention shall be those in force at the time a license was issued or renewed.

14.4. All fees and payments to the Authority shall be transmitted in convertible currency.

APPENDIX B

Terms and Procedures Applying to Licenses in the International Seabed Area Beyond the International Trusteeship Area

1. Entities Entitled to Obtain Licenses:

1.1. Contracting Parties or a group of Contracting Parties, one of which shall act as the operating or sponsoring Party for purposes of fixing operational or supervisory responsibility, are authorized to apply for and obtain exploration and exploitation licenses. Any Contracting Party or group of Contracting Parties, which applies for a license to engage directly in exploration or exploitation, shall designate a specific agency to act as operator on its behalf for the purposes of this Convention.

1.2. Natural or juridical persons are authorized to apply for and obtain exploration and exploitation licenses from the International Seabed Resource Authority if they are sponsored by a Contracting Party.

2. Exploration Licenses—Procedures:

2.1. Licenses shall be issued promptly by the Authority through the Sponsoring Party to applicants meeting the requirements specified in Appendix A.

3. Exploitation Licenses—Procedures:

3.1. The Sponsoring Party shall certify as to the technical and financial competence of the operator, and shall transmit the operator's work plan.

3.2. An application for an exploitation license shall be preceded by a notice of intent to apply for a license submitted by the operator to the Authority and the prospective Sponsoring Party. Such a notice of intent, when accompanied by evidence of the deposit of the license fee referred to in paragraph 4.1 of Appendix A, shall reserve the block for one hundred and eighty days. Notices of intent may not be renewed.

3.3. Notices of intent shall be submitted sealed to the Authority and opened at monthly intervals at previously announced times.

3.4. Subject to compliance with these procedures, if only one notice of intent has been received for a particular block, the applicant shall be granted a license, except as provided in paragraphs 3.6. through 3.8.

3.5. If more than one notice of intent to apply for a license for the same block or portion thereof is received at the same opening, the Authority shall notify the applicants and their Sponsoring Parties that the exploitation license to the block or portion thereof will be sold to the highest bidder at a sale to be held one hundred and eighty days later, under the following terms:

(a) The bidding shall be on a cash bonus basis and the minimum bid shall be twice the license fee;

(b) Bids shall be sealed;

(c) The bidding shall be limited to such of the original applicants whose applications have been received in the interim from their sponsoring Parties;

(d) Bids shall be announced publicly by the Authority when they are opened. In the event of a tie, the tie bidders shall submit a second sealed bid to be opened 28 days later;

(e) The final award shall be announced publicly by the Authority within seven days after the bids have been opened.

3.6. In the event of the termination, forfeiture, or revocation of an exploitation license to a block, or relinquishment of a part of a block, the block or portion thereof will be offered for sale by sealed competitive bidding on a cash bonus basis in addition to the current license fee. The following provisions shall apply to such a sale:

(a) The availability of such a block, or portion thereof, for bidding shall be publicly announced by the Authority as soon as possible after it becomes available, and a sale following the above procedures shall be held within one hundred and eighty days after a request for an exploitation license on the block has been received;

(b) The bidding shall be opened to all sponsored operators, including, except in the case of revocation, the operator who previously held the exploitation license to the block or to the available portion thereof;

(c) If the winning bid is submitted by an operator who previously held the exploitation right to the same block, or to the same portion thereof, the work requirement will begin at the level that would have applied if the operator had continuously held the block.

3.7. Blocks, or portions thereof, contiguous to a block on which production has begun shall also be sold by sealed competitive bidding under the terms specified in paragraph 3.6.

3.8. Blocks, or separate portions thereof, from which hydrocarbons or other fluids are being drained, or are believed to be drained, by production from another block shall be offered for sale by sealed competitive bidding under the terms specified in paragraph 3.6 at the initiative of the Authority.

3.9. Geologic and other data concerning blocks, or portions thereof, open for bidding pursuant to paragraphs 3.6 through 3.8, which are no longer confidential, shall be made available to the public prior to the bidding date. Data on blocks, or separate portions thereof, for which the license has been revoked for violations shall be made available to the public within 30 days after revocation.

3.10. Exploitation licenses shall only be transferable with the approval of the Sponsoring Party and the Authority, provided that the transferee meets the requirements of this Convention, is sponsored by a Contracting Party, and a transfer fee is payed to the Authority in the amount of $250,000. This fee shall not apply in transfers between parts of the same operating enterprise.

4. Duration of Exploitation Licenses:

4.1. If commercial production has been achieved within fifteen years after the license has been issued, the exploitation license shall be extended automatically for twenty additional years from the data commercial production has commenced.

4.2. At the completion of the twenty-year production period referred to in paragraph 4.1, the operator with the approval of the Sponsoring Party shall have the option to renew his license for another twenty years at the rental fees and payment rates in effect at the time of renewal.

4.3. At the end of the forty-year term, or earlier if the license is voluntarily relinquished or expires pursuant to paragraph 5.9 of Appendix A, the block or blocks, or separate portions of blocks, to which the license applied shall be offered for sale by competitive bidding on a cash bonus basis. The previous licensee shall have no preferential right to such block, or separate portion thereof.

5. Work Requirements:

5.1. The annual work requirement fee per block shall be specified in an Annex in accordance with the following schedule:

Paragraph 5.1 (a) and (c) minerals

Years:	Amounts per annum
1 to 5	$20,000 to $60,000.
6 to 10	$180,000 to $540,000.
11 to 15	$200,000 to $600,000.
Total	$2,000,000 to $6,000,000.

Paragraph 5.1 (b) minerals

Years:	Amounts per annum
1 to 2	$20,000 to $60,000.
3 to 10	$120,000 to $360,000.
11 to 15	$200,000 to $600,000.
Total	$2,000,000 to $6,000,000.

The minimum annual work requirement fee for a portion of a block shall be an appropriate fraction of the above, to be specified in an Annex.

5.2. Work expenditures with respect to one or more blocks may be considered as meeting the aggregate work requirements on a group of blocks originally licensed in the same year, to the same operator, in the same category, provided that the number of such blocks shall not exceed sixteen in the case of the category of minerals described in paragraph 5.1(a) of Appendix A, four in the case of the category of minerals described in paragraph 5.1(b) of Appendix A and eight in the case of the category of minerals described in paragraph 5.1(c) of Appendix A.

5.3. Should the aggregate work requirement fee of $2,000,00 to $6,000,000 be spent prior to the end of the thirteenth year, an additional work requirement fee of $25,000-$50,000, as specified in an Annex, shall be met until commercial production begins or until expiration of the fifteen-year period.

5.4. After commercial production begins, the operator shall at the beginning of each year deposit $100,000 to $200,000 as specified in an Annex, or with the Sponsoring Party post a bond for that amount. Such deposit or bond shall be returned in an amount equivalent to expenditures on or related to the block and the value of production at the site. A portion [a figure between 50% and 66 2/3% will be inserted here] of any funds not returned shall be transmitted to the Authority.

6. Unit Management:

The Operations Commission shall assure unit management and production pursuant to Section 9 of Appendix A, giving the licensees and their Sponsoring Parties a reasonable time to reach agreement on a plan for unit operation.

APPENDIX C

Terms and Procedures for Licenses in the International Trusteeship Area

1. General:

1.1. Unless otherwise specified in this Convention, all provisions of this Convention except those in Appendix B shall apply to the International Trusteeship Area.

2. Entities Entitled to Obtain Licenses:

2.1. The Trustee Party, pursuant to Chapter III, shall have the exclusive right, in its discretion, to approve or disapprove applications for exploration and exploitation licenses.

3. Exploration and Exploitation Licenses:

3.1. The Trustee Party may use any system for issuing and allocating exploration and exploitation licenses.

3.2. Copies of licenses issued shall be forwarded to the Authority.

4. Categories and Size of Blocks:

4.1. The Trustee Party may license separately one or more related minerals of the categories listed in paragraph 5.1 of Appendix A.

4.2. The Trustee Party may establish the size of the block for which exploitation licenses are issued within the maximum limits specified in Appendix A.

5. Duration of Exploitation Licenses:

5.1. The Trustee Party may establish the term of the exploitation license and the conditions, if any, under which it may be renewed, provided that its continuance after the first 15 years is contingent upon the achievement of commercial production.

6. Work Requirements:

6.1. The Trustee Party may set the work requirements at or above those specified in Appendix A and put these in terms of work to be done rather than funds to be expended.

7. Unit Management:

7.1. When a deposit most rationally extracted under unit management lies wholly within the International Trusteeship Area, or astride its landward boundary, the Trustee Party concerned shall assure unit management and production pursuant to Section 9.1 of Appendix A, and shall submit the plan for unit operation to the Operations Commission.

7.2. With respect to deposits lying astride a boundary between two Trustee Parties in the International Trusteeship Area, such Parties shall agree on a plan to assure unit management and production, and shall submit the operation plan to the Operations Commission.

8. Proration:

8.1. The Trustee Party may establish proration, to the extent permitted by its domestic law.

9. Payments:

9.1. Pursuant to Subparagraph (e) of Article 28, the Trustee Party may collect fees and payments related to the issuance or retention of a license in addition to those specified in this Convention, including but not limited to payments on production higher than those required by this Convention.

9.2. The Trustee Party shall transfer to the Authority a portion [a figure between 50% and 66 2/3% will be inserted here] of the fees and payments referred to in paragraph 9.1 except as otherwise provided in paragraphs 3.3, 4.4 and 6.4 of Appendix A.

Note—Further study is required on the means to assure equitable application of the principle contained in paragraph 9.2 to socialist and non-socialist parties and their operations.

10. Standards:

10.1. The Trustee Party may impose higher operating, conservation, pollution, and safety standards than those established by the Authority, and may impose additional sanctions in case of violence of applicable standards.

11. Revocation:

11.1. The Trustee Party may suspend or revoke licenses for violation of this Convention, or of the rules it has established pursuant thereto, or in accordance with the terms of the license.

APPENDIX D

Division of Revenue

1. Disbursements:

1.1. All disbursements shall be made out of the net income of the Authority, except as otherwise provided in paragraph 2 of Article 74.

2. Administrative Expenses of the International Seabed Resource Authority:

2.1. The Council, in submitting the proposed budget to the Assembly, shall specify what proportion of the revenues of the Authority shall be used for the payment of the administrative expenses of the Authority.

2.2. Upon approval of the budget by the Assembly, the Secretary-General is authorized to use the sums allotted in the budget for the expenses specified therein.

3. Distribution of the Net Income of the Authority:

3.1. The net income, after administrative expenses, of the Authority shall be used to promote the economic advancement of developing States Parties to this Convention and for the purposes specified in paragraph 2 of Article 5 and in other Articles of this Convention.

3.2. The portion to be devoted to economic advancement of developing States Parties to this Convention shall be divided among the following international development organizations as follows:

Note.—A list of international and regional development organizations should be included here, indicating percentages assigned to each organization.

3.3. The Council shall submit to the Assembly proposals for the allocation of the income of the Authority within the limits prescribed by this Appendix.

3.4. Upon approval of the allocation by the Assembly, the Secretary-General is authorized to distribute the funds.

APPENDIX E

Designated Members of the Council

1. Those six Contracting Parties which are both developed States and have the highest gross national product shall be considered as the six most industrially advanced Contracting Parties.

2. The six most industrially advanced Contracting Parties at the time of the entry into force of this Convention shall be deemed to be:

________________________________.

They shall hold office until replaced in accordance with this Appendix.

3. The Council, prior to every regular session of the Assembly, shall decide which are the six most industrially advanced Contracting Parties. It shall make rules to ensure that all questions relating to the determination of such Contracting Parties are considered by an impartial committee before being decided upon by the Council.

4. The Council shall report its decision to the Assembly, together with the recommendations of the impartial committee.

5. Any replacements of the designated members of the Council shall take effect on the day following the last day of the Assembly to which such a report is made.

APPENDIX L

UNITED STATES REVISED DRAFT FISHERIES ARTICLE*
(August 1972)

I. REGULATORY AUTHORITY

Authority to regulate the living resources of the high seas shall be determined by their biological characteristics and such authority shall be exercised so as to assure their conservation, maximum utilization and equitable allocation.

II. COASTAL AND ANADROMOUS LIVING RESOURCES

The coastal State shall regulate and have preferential rights to all coastal living resources off its coast beyond the territorial sea to the limits of their migratory range. The coastal State in whose fresh or estuarine waters anadromous resources (e.g. salmon) spawn shall have authority to regulate and have preferential rights to such resources beyond the territorial sea throughout their migratory range on the high seas (without regard to whether or not they are off the coast of said State).

A. The term "coastal resource" refers to all living resources off the coast of a coastal State except the highly migratory species listed in Annex A[a] and anadromous resources.

B. The coastal State may annually reserve to its flag vessels, in accordance with this article, that portion of such coastal and anadromous resources as they can harvest.

C. Such coastal and anadromous resources which are located in or migrate through waters adjacent to more than one coastal State shall be regulated by agreement among such States.

---

*Originally issued as document A/AC.138/SC.II/L.9.

[a]Annex A not attached.

III. HIGHLY MIGRATORY OCEANIC RESOURCES

The highly migratory oceanic resources listed in Annex A shall be regulated by appropriate international fishery organizations.

A. Any coastal State party, or other State party whose flag vessels harvest or intend to harvest a regulated resource, shall have an equal right to participate in such organizations.

B. No State party whose flag vessels harvest a regulated resource may refuse to co-operate with such organizations. Regulations of such organizations in accordance with this Article shall apply to all vessels fishing the regulated resources regardless of their nationality.

C. In the event the States concerned are unable or deem it unnecessary to establish an international organization the resources shall be regulated by agreement or consultation among such States.

IV. CONSERVATION PRINCIPLES

In order to assure the conservation of living marine resources, the coastal State or appropriate international organization shall apply the following principles:

A. Allowable catch and other conservation measures shall be established which are designed, on the basis of the best evidence available, to maintain or restore the maximum sustainable yield, taking into account relevant environmental and economic factors.

B. For this purpose scientific information, catch and effort statistics, and other relevant data shall be contributed and exchanged on a regular basis.

C. Conservation measures and their implementation shall not discriminate in form or fact against any fishermen. Conservation measures shall remain in force pending the settlement, in accordance with the relevant provisions of this Article, of any disagreement as to their validity.

V. UTILIZATION AND ALLOCATION

In order to assure the maximum utilization and equitable allocation of coastal and anadromous resources, the coastal State shall apply the following principles:

A. The coastal State may reserve to its flag vessels that portion of the allowable annual catch they can harvest.

B. The coastal State shall provide access by other states, under reasonable conditions, to that portion of the resources not fully utilized by its vessels on the basis of the following priorities:

(1) States that have traditionally fished for a resource, subject to the conditions of sub-paragraph C;

(2) other States in the region, particularly landlocked States and other States with limited access to the resources, with whom joint or reciprocal arrangements have been made; and

(3) all States, without discrimination among them.

C. Whenever necessary to accommodate the allocations to the coastal States traditional fishing may be reduced, without discrimination among those States that have traditionally fished for a resource, in the following manner:

(Formula to be negotiated within Subcommittee II which takes into account the interests of traditional fishing States.)

States whose fishermen harvest a resource under regulation by a coastal State may be required, without discrimination, to pay reasonable fees to defray their share of the cost of such regulation.

VI. NOTIFICATION CONSULTATION

The coastal State shall give to all affected states timely notice of any conservation, utilization and allocation regulations, prior to their implementation, and shall consult with other States concerned.

VII. TECHNICAL ASSISTANCE

An international register of independent fisheries experts shall be established and maintained by the Food and Agriculture Organization of the United Nations.[b] Any developing State party to this convention requiring assistance may select an appropriate number of such experts to serve as a fishery management advisory group to that state.

VIII. ENFORCEMENT

Actions under this paragraph shall be taken in such a manner as to minimize interference with fishing and other activities in the

---

[b]The Sub-Committee may wish, in accordance with paragraph 13 of General Assembly Resolution 2750C (XXV), to invite the comments of the Director-General on the Food and Agriculture Organization of the United Nations on the ability of the Organization to assume such responsibilities.

marine environment.

A. Coastal State - the coastal State may inspect and arrest vessels for fishing in violation of its regulations. The coastal State may try and punish vessels for fishing in violation of its regulations, provided that where the state of nationality of a vessel has established procedures for the trial and punishment of violations of coastal State fishing regulations adopted in accordance with this article, an arrested vessel shall be delivered promptly to duly authorized officials of the State of nationality for trial and punishment, who shall notify the coastal state of the disposition of the case within six months.

B. International fisheries organization - Each State party to an international organization shall make it an offence for its flag vessels to violate the regulations adopted by such organization in accordance with this article. Officials authorized by the appropriate international organization, or of any State so authorized by the organization, may inspect and arrest vessels for violating the fishery regulations adopted by such organizations. An arrested vessel shall be promptly delivered to the duly authorized officials of the flag State. Only the flag State of the offending vessel shall have jurisdiction to try the case or impose any penalties regarding the violation of fishery regulations adopted by international organizations pursuant to this article. Such State has the responsibility of notifying the enforcing organization within a period of six months of the disposition of the case.

IX. DISPUTES SETTLEMENT

Any dispute which may arise between States under this article shall, at the request of any of the parties to the dispute, be submitted to a special commission of five members unless the parties agree to seek a solution by another method of peaceful settlement, as provided for in Article 33 of the Charter of the United Nations. The commission shall proceed in accordance with the following provisions.

A. The members of the commission, one of whom shall be designated as chairman, shall be named by agreement between the States in dispute within two months of the request for settlement in accordance with the provisions of this article. Failing agreement they shall, upon request of any State party to the dispute, be named by the Secretary General of the United Nations, within a further two-month period, in consultation with the States involved and with the President of the International Court of Justice and the Director-General of the Food and Agriculture Organization of the United Nations, from amongst well-qualified persons being nationals of States not involved in the dispute and specializing in legal, administrative or scientific questions relating to fisheries, depending upon the nature of the dispute to be settled. Any vacancy arising after the original appointment shall be filled in the same manner as provided for the initial selection.

B. Any State party to proceedings under these articles shall have the right to name one of the nationals to sit with the special commission, with the right to participate fully in the proceedings on the same footing as a member of the commission but without the right to vote or to take part in the writing of the commission's decision.

C. The commission shall determine its own procedure, assuring each party to the proceedings a full opportunity to be heard and to present its case. It shall also determine how the costs and expenses shall be divided between the parties to the dispute failing agreement by the parties on this matter.

D. Pending the final award by the special commission, measures in dispute relating to conservation shall be applied; the commission may decide whether and to what extent other measures shall be applied pending its final award.

E. The special commission shall render its decision, which shall be binding upon the parties, within a period of five months from the time it is appointed unless it decides, in the case of necessity to extend the time limit for a period not exceeding two months.

F. The special commission shall, in reaching its decision, adhere to this article and to any agreements between the disputing parties implementing this article.

X. OTHER USES

The exploitation of the living resources shall be conducted with reasonable regard for other activities in the marine environment.

XI. EXISTING CONVENTIONS

The provisions of this article may be applied to fishery conventions and other international fishery agreements already in force.

LIVING RESOURCES CLAIMED BY THE UNITED STATES AS RESOURCES OF THE CONTINENTAL SHELF IN ACCORDANCE WITH THE CONTINENTAL SHELF CONVENTION

Sponges

Glove Sponge—Hippiospongia canaliculata;
Sheepswool Sponge—Hippiospongia lachne;
Grass Sponge—Spongia graminea;
Yellow Sponge—Spongia barbera.

Crustacea

Tanner Crab—Chionoecetes tanneri;
Tanner Crab—Chionoecetes opilio;
Tanner Crab—Chionoecetes angulatus;
Tanner Crab—Chionoecetes bairdi;
King Crab—Paralithodes camtschatica;
King Crab—Paralithodes platypus;
King Crab—Paralithodes brevipes;
Stone Crab—Menippe mercenaria;
Lobster—Homarus Americanus;
Dungeness Crab—Cancer magister;
California King Crab—Paralithodes californiensis;
Golden King Crab—Lithodes aequispinus;
Northern Stone Crab—Lithodes maia;
Stone Crab—Menippe mercenaria; and
Deep-Sea Red Crab—Ceryon quinquedens.

Note: With the exception of lobster (Homarus Americanus) all the above species have been placed on this list by administrative action. Lobster was added by act of Congress.

Source: Public Law 93-242, section 15.

Mollusks

Red Abalone—Haliotis rufescens;
Pink Abalone—Haliotis corrugata;
Japanese Abalone—Haliotis kamtschatkana;
Queen Conch—Strombus gigas;
Surf Clam—Spisula solidissima; and
Ocean Quahog—Artica islandica.

UNITED STATES DRAFT ARTICLES FOR A CHAPTER ON THE RIGHTS AND DUTIES OF STATES IN THE COASTAL SEABED ECONOMIC ZONE

Article 1

1. The coastal State shall have the exclusive right to explore and exploit and authorize the exploration and exploitation of the natural resources of the seabed and subsoil in accordance with its own laws and regulations in the Coastal Seabed Economic Area.

2. The Coastal Seabed Economic Area is the area of the seabed which is

(a) seaward of—; and

(b) landward of an outer boundary of—.

3. The coastal State shall in addition have the exclusive right to authorize and regulate in the Coastal Seabed Economic Area or the superjacent waters:

(a) the construction, operation and use of offshore installations affecting its economic interests, and

(b) drilling for purposes other than exploration and exploitation of resources.

4. The coastal State may, where necessary, establish reasonable safety zones around such offshore installations in which it may take appropriate measures to protect persons, property, and the marine environment. Such safety zones shall be designed to ensure that they are reasonably related to the nature and function of the installation. The breadth of the safety zones shall be determined by the coastal State and shall conform to international standards in existence or to be established pursuant to Article 3.

5. (a) For the purposes of this Chapter, the term "installations" refers to all offshore facilities, installations, or devices other than those which are mobile in their normal mode of operation at sea.

This Chapter deals with seabeds resources, and does not deal with fisheries. The proposal of the United States with respect to fisheries beyond the territorial sea was introduced in Subcommittee II on August 4, 1972 (A/AC.138/Sc.II/SR.40) (Official Record of the General Assembly, Twenty-seventh session, Supplement No. 21, A/8721).

(b) Installations do not possess the status of islands. They have no territorial sea or Coastal Seabed Economic Area of their own, and their presence does not affect the delimination of the territorial sea of the coastal State.

6. The coastal State may, with respect to the activities set forth in this Article, apply standards for the protection of the marine environment higher than those required by applicable international standards pursuant to Article 2.

7. The coastal State may, with respect to the activities set forth in this Article, take all necessary measures to ensure compliance with its laws and regulations subject to the provisions of this Chapter.

Article 2

The coastal State, in exercising the rights referred to in Article 1, shall ensure that its laws and regulations, and any other actions it takes pursuant thereto in the Coastal Seabed Economic Area, are in strict conformity with the provisions of this Chapter and other applicable provisions of this Convention, and in particular:

(a) the coastal State shall ensure that there is no unjustifiable interference with other activities in the marine environment, and shall ensure compliance with international standards in existence or promulgated by the Authority or the Inter-Governmental Maritime Consultative Organization, as appropriate, to prevent such interference;

(b) the coastal State shall take appropriate measures to prevent pollution of the marine environment from the activities set forth in Article 1 and shall ensure compliance with international standards in existence or promulgated by the Authority or the Inter-Governmental Maritime Consultative Organization, as appropriate, to prevent such pollution;

(c) the coastal State shall not impede, and shall co-operate with the Authority in the exercise of its inspection functions in connection with subparagraph (b) above;

(d) the coastal State shall ensure that licenses, leases, or other contractual arrangements which it enters into with the agencies of instrumentalities of other States, or with natural or juridical persons which are not nationals of the coastal State, for the purpose of exploring for or exploiting seabed resources are strictly observed according to their terms. Property of such agencies, instrumentalities or persons shall not be taken except for a public purpose, on a non-discriminatory basis, nor shall it be taken without the prompt payment of just compensation. Such compensation shall be in an effectively realizable form and shall represent the full equivalent of the property taken and adequate provision shall have been made at or prior to the time of the taking to ensure compliance with the provisions of this paragraph;

(e) the coastal State shall make available in accordance with the provisions of Article—, such share of revenues in respect of mineral resources exploitation from such part of the Coastal Seabed Economic Area as is specified in that Article.

Article 3

1. All activities in the marine environment shall be conducted with reasonable regard to the rights of the coastal State referred to in Article 1.

2. States shall ensure compliance with international standards in existence or to be promulgated by Inter-Government Maritime Consultative Organization in consultation with the Authority:

(a) regarding the breadth, if any, of safety zones around offshore installations;

(b) regarding navigation outside the safety zones, but in the vicinity of offshore installations.

Article 4[2]

Nothing in this Chapter shall affect the rights of freedom of navigation and overflight and other rights to carry on activities unrelated to seabed resource exploration and exploitation in accordance with general principles of international law, except as otherwise specifically provided in this Convention.

Article 5

Any dispute with respect to the interpretation or application of the provisions of this Chapter shall, if requested by either party to the dispute, be resolved by the compulsory dispute settlement procedures contained in Article—, of Chapter—.

---

2. It is assumed that the general articles of the Law of the Sea Convention will contain an article such as Article 4 applicable to all areas beyond the territorial sea. Such an article would obviate the need for several articles making the same point here and in other chapters of the Convention.

APPENDIX O

UNITED STATES DRAFT ARTICLES ON THE PROTECTION OF THE MARINE ENVIRONMENT AND THE PREVENTION OF MARINE POLLUTION

Section A: Obligations to Protect the Marine Environment

Article I: General Obligation

Article II: Particular Obligations

(Two articles on these subjects were discussed during the March/April meeting of the Seabed Committee. We take note of those drafts and the footnotes and will of course, participate in the latter consideration of them in the Working Group and the Subcommittee.)

Section B: Competence to Establish Standards to Protect the Marine Environment

Article III: International Standards in General

1. The Authority established by Chapter — of this Convention shall have primary responsibility for establishing, as soon as possible and to the extent they are not in existence, international standards with respect to the seabeds.

2. The Intergovernmental Maritime Consultative Organization shall have primary responsibility for establishing, as soon as possible and to the extent they are not in existence, international standards with respect to vessels.

3. Such standards may include special standards for special areas and problems, taking into account particular ecological circumstances.

4. These organizations shall cooperate with each other, other international organizations in the field, and the United Nations Environment Program.

Article IV: The Right and Duty to Implement Standards

States shall adopt laws and regulations implementing international standards in respect of marine based sources of pollution of the marine environment or may adopt and implement higher standards:

(a) in the exercise of their rights in the [Coastal Seabed Economic Area] with respect to the activities set forth in Chapter—, Article—of this Convention;

(b) for vessels entering their ports and offshore facilities;

(c) for their nationals, natural or juridical, and vessels registered in their territory or flying their flag.

Section C: General Competence to Enforce Standards to Protect the Marine Environment

Article V: Enforcement Instrumentalities

For the purposes of this Chapter, a State shall act through duly authorized government vessels, aircraft, or officials. Any State may, by agreement, authorize one or more other States to act for it in taking pollution enforcement measures and shall so inform other States through IMCO, or directly.

Article VI: Enforcement in the [Coastal Seabed Economic Area]

1. In the exercise of its rights in the [Coastal Seabed Economic Area] pursuant to Chapter—, the coastal State shall enforce the standards applicable in accordance with the provisions of this Chapter to the activities set forth in Chapter —, Article—of this Convention.

2. The Authority established in Chapter—, may inspect, in accordance with Article—, the activities specified in paragraph one of this Article, in cooperation with the coastal State, to ensure that the activities are being conducted in compliance with the standards applicable in accordance with the provisions of this Chapter.

Article VII: Ordinary Enforcement Against Vessels

1. A State shall enforce standards applicable in accordance with the provisions of this Chapter to vessels registered in its territory or flying its flag (such State is hereinafter referred to as the "flag State").

2. A State may enforce standards applicable in accordance with the provisions of this Chapter to:

(a) vessels using its ports or offshore facilities irrespective of where the violation occurred, provided, however, that such proceedings are commenced no later than [three years] after such violation occurred (such State is hereinafter referred to as the "port State").

(b) vessels in its territorial sea for violations therein, except as otherwise provided in this Convention.

Section D: Cooperative Enforcement Measures Against Vessels

Article VIII: The Right to Monitor

A vessel within or beyond the territorial sea shall upon request by any duly authorized government vessel, aircraft or official in the vicinity which has reason to suspect a violation of the applicable international standards, give information specifying its name, State of registry, next scheduled ports of call, and any other information required to be given by the applicable international standards.

Article IX: Denial of Port Entry

Any State may inform a vessel at any time that it will be denied entry to its ports for non-compliance with any of its environmental requirements or its refusal to allow an immediate on-board inspection to determine the source of possible pollution. Any State may, by agreement, authorize one or more other States to act for it in this respect and shall so inform other States through IMCO, or directly.

Article X: The Duty to Notify

If a State has reason to suspect a violation of the applicable international standards, it shall notify the flag State or the State of one of the next ports of call or both, of the alleged violation and forward the available evidence.

Article XI: Port State Duties

Upon receipt of such notification of the alleged violation, the port State shall undertake, upon arrival of the vessel if within six months of the alleged violation, an immediate and thorough investigation. The port State shall promptly inform the flag State and the notifying State of the results of the investigation and its actions including a statement as to whether it intends to institute proceedings and the result of any such proceedings.

Article XII: Flag State Duties

Upon receipt of notification if within six months of an alleged violation, the flag State shall undertake an immediate and thorough investigation. If the result of its or a port State's investigation indicates that a violation has occurred, the flag State shall institute proceedings against the vessel, its operator, its master, or its owner, provided that it shall not be required to do so if proceedings have already taken place in respect of violation. The flag State shall inform the notifying State and any other State which could institute proceedings of its decisions and actions.

Article XIII: Participation in Investigations

A notifying State may participate in any investigation undertaken pursuant to its notification. A flag State may designate an observer for any investigation involving one of its flag vessels. An expert or experts designated by IMCO shall be permitted to participate in any investigation if so requested by a State concerned and such expert or experts may file a separate report with IMCO.

Section E: Extraordinary Enforcement Measures and Intervention Against Vessels

Article XIV: Coastal State Remedy Against Flag States

If the dispute settlement machinery established in Chapter— finds, upon petition by any State, that a particular flag State has unreasonably and persistently failed to enforce the applicable international standards against its flag vessels, the machinery may specify additional enforcement measures which may be taken by coastal States for violations by any vessel of that flag. Such authorization shall be interim in nature and shall be limited to those measures necessary to bring about adequate flag State enforcement. Such authorization shall be rescinded upon a showing by the flag State that it is taking adequate measures.

Article XV: Emergency Coastal State Enforcement Procedures[3]

Beyond the territorial sea, a coastal State may take such reasonable emergency enforcement measures as may be necessary to prevent, mitigate or eliminate imminent danger of major harmful damage to its coast or related interests from pollution arising from a particular occurrence reasonably believed to be related to a violation of the applicable international standards.

Article XVI: Intervention

(The 1969 Intervention Convention allows coastal States to take such measures on the high seas as may be necessary to prevent, mitigate or eliminate grave and imminent danger to their coastline or related interests from pollution or threat of pollution by oil following upon a maritime casualty or acts related to such casualty, which may reasonably be expected to result in major harmful consequences. Negotiations are presently under way in IMCO to expand the Convention to cover other substances in addition to oil and the issue is on the agenda for the October 1973 Conference on Marine Pollution. Consequently, it may be advisable to await the results of these negotiations before taking action on the issue in the law of the sea negotiations.)

Section F: General Articles Relating to Enforcement

Article XVII: Release of Vessels

A vessel shall be permitted to continue its voyage and shall not be detained longer than its presence is essential for investigative

---

3. It should be noted that this differs from and is in addition to the measures set out in the Convention relating to Intervention on the High Seas in Cases of Oil Pollution Casualties.

purposes. It shall be promptly released if the investigation does not reveal a violation of the standards applicable in accordance with the provisions of this Chapter. Where there continues to be reason to believe a violation has occurred, vessels shall be promptly released under reasonable procedures such as bonding except where such release would present an unreasonable threat of harm to the marine environment or where other action is required or authorized by the applicable international standards.

Article XVIII: Penalties

All violations of the applicable international standards shall be prohibited under the law of each State. The penalties provided for such violations shall be applied so as to guarantee fair treatment, shall be adequate in severity to discourage any such violation, and shall, in any case, be at least as severe as those applied by that State for violations in its territorial sea.

Article XIX: Multiple Proceedings

Whenever a State other than the flag State has instituted proceedings against a vessel, its operator, its master, or its owner, no other proceedings in respect of the same violation may be instituted except by the flag State of the vessel or by any other State in whose territorial sea or internal waters the violation has taken place. In assessing penalties, a State shall take into account any penalties assessed by other States in respect to the same violation. This shall not restrict the right of any State or person to institute a suit or claim for damages caused by pollution.

Article XX: Cooperation

States shall afford one another the greatest measure of assistance in carrying out the objectives of this Chapter and in particular in providing evidence and witnesses necessary for investigations and proceedings.

Section G: Other Articles Relating to the Marine Environment

Article XXI: Liability for Unreasonable Measures

A State shall be liable for damage resulting from investigative, enforcement or intervention measures exceeding those reasonably necessary in the light of available information.

Article XXII: State Responsibility

1. A State has the responsibility to take appropriate measures to ensure, in accordance with international law, that activities under its jurisdiction or control do not cause damage to the environment of other States or to the marine environment beyond the limits of national jurisdiction.

2. States shall undertake, as soon as possible, jointly to develop international law regarding liability and compensation for pollution damage including, inter alia, procedures and criteria for the determination of liability, the limits of liability and available defenses.

3. In the absence of other adequate remedies with respect to damage to the environment of other States caused by activities under the jurisdiction or control of a State, that State has the responsibility to provide recourse for foreign states or nationals to a domestic forum empowered:

(a) to require the abatement of a continuing source of pollution of the marine environment, and

(b) to award compensation for damages.

Article XXIII: Sovereign Immunity

This Chapter shall not apply to those vessels and aircraft entitled to sovereign immunity under international law. However, each State shall ensure, by the adoption of appropriate measures, that all such vessels and aircraft owned or operated by it act in a manner consistent with the object and purpose of this Chapter.

Article XXIV: Dispute Settlement

Any dispute with respect to the interpretation or application of the provisions of this Chapter shall, if requested by any party to the dispute, be resolved by the compulsory dispute settlement procedures contained in Chapter—.

APPENDIX P

UNITED STATES DRAFT ARTICLES FOR A CHAPTER ON MARINE SCIENTIFIC RESEARCH

Article 1

Scientific research in the sea being essential to an understanding of global environment, the preservation and enhancement of the sea and its rational and effective use, States shall promote and facilitate the development and conduct of all scientific research in the sea for the benefit of the international community. All States, irrespective of geographical location, as well as appropriate international organizations may engage in scientific research in the sea, recognizing the rights and interests of the international community and coastal States, particularly the interests and needs of developing countries, as provided for in this Convention.

Article 2

Scientific research shall be conducted with reasonable regard to other uses of the sea, and such other uses shall be conducted with reasonable regard to the conduct of scientific research.[1]

Article 3

Scientific research shall be conducted with strict and adequate safeguards for the protection of the marine environment[2]

---

1. A general treaty article on the subject of accommodation of uses dealing with all uses of the sea might be included in the general articles of the Law of the Sea Convention. This could obviate the need for a specific article for each use, such as that suggested above for scientific research, that prepared as Text 17 of the texts dealing with principles for the seabed areas beyond the limits of national jurisdiction (A/AC.138/SC.1/L.22, Apr. 4, 1973) or that included in art. 2, ¶ 2, of the Convention on the High Seas.

2. The need for a wording of the article might be further considered in the light of the draft articles prepared by the Working Group on marine pollution.

Article 4

Scientific research activities shall not form the legal basis for any claim to any part of the sea or its resources.[3]

Article 5

States shall promote international cooperation in scientific research exclusively for peaceful purposes:

(a) by participating in international programs and by encouraging cooperation in scientific research by personnel of different countries;

(b) through effective publication of scientific research programs and dissemination of the results of such research through international channels and promotion of the flow of scientific research to developing countries;

(c) through measures to strengthen scientific research capabilities of developing countries, including assistance in assessing the implications for their interests of scientific research data and results, the participation of their nationals in research programs, and education and training of their personnel.

Article 6

Coastal States in the exercise of their sovereignty shall cooperate in facilitating the conduct of scientific research in their territorial sea and access to their ports by research vessels.

Article 7

In areas beyond the territorial sea where the coastal State exercises jurisdiction pursuant to Articles—over seabed resources and coastal fisheries, States and appropriate international organizations

---

3. There may be merit in the inclusion of an article in the general articles of the Law of the Sea Convention to the effect that no claims to any part of the sea can be made except as specifically provided in the Convention. This could obviate the need for a specific article for each use of the sea (see e.g. Texts 4 and 11 of the draft seabed articles prepared by the Working Group of Subcommittee I: Article 2 of the U.S. draft seabeds treaty).

shall ensure that their vessels conducting scientific research shall respect the rights and interests of the coastal State in its exercise of such jurisdiction, and for this purpose shall:

(a) provide the coastal State at least—days advance notification of intent to do such research, containing a description of the research, containing a description of the research project which shall be kept up to date;

(b) certify that the research will be conducted in accordance with this Convention by a qualified institution with a view of purely scientific research;

(c) ensure that the coastal State has all appropriate opportunities to participate or be represented in the research project directly or through an appropriate international institution of its choice; the coastal State shall give reasonable advance notification of its desire to participate or be represented in the research within—days after it has received notification;

(d) ensure that all data and samples are shared with the coastal State;

(e) ensure that significant research results are published as soon as possible in an open readily available scientific publication and supplied directly to the coastal State;

(f) assist the coastal State in assessing the implications for its interests of the data and results directly or through the procedures established pursuant to Article 5;

(g) ensure compliance with all applicable international environmental standards, including those established or to be established by [insert name or names of appropriate organizations].

Article 8

Any dispute with respect to the interpretation or application of the provisions of this Chapter shall, if requested by either party to the dispute, be resolved by the compulsory dispute settlement procedures contained in Article—.

UNITED STATES DRAFT ARTICLES FOR A CHAPTER ON THE SETTLEMENT OF DISPUTES

Article 1

In any dispute between the Contracting Parties relating to the interpretation or application of the present Convention, any party to the dispute may invite the other party or parties to the dispute to settle the dispute by direct negotiation, good offices, mediation, conciliation, arbitration, or through special procedures provided for by an international or regional organization.

Article 2

Notwithstanding the provisions of Article 1, any Contracting Party which is a party to a dispute relating to the interpretation or application of this Convention which is required by this Convention to be submitted to compulsory dispute settlement procedures on the application of one of the parties, may refer the dispute at any time to the Law of the Sea Tribunal (the Tribunal).

Article 3

Notwithstanding the provisions of Article 2, if the parties to a dispute have agreed in any general, regional, or special agreement to resort to arbitration, any party to the dispute shall be entitled to refer it to arbitration in accordance with that agreement in place of the procedures specified in this Chapter.

Article 4

The Tribunal shall be established and shall function in accordance with the annexed Statute. Its members shall be nominated and elected in accordance with the procedure provided for in the Statute of the International Court of Justice for the election of the judges of the Court. They shall be lawyers of recognized competence in law of the sea matters.

Article 5

When a dispute involves technical questions, such as safety of navigation, ship construction, pollution, scientific research, fishing, or seabed exploration or exploitation, the Tribunal shall be assisted in the consideration of the case by four technical assessors sitting with it, but without the right to vote. These assessors shall be chosen by the President of the Tribunal from the list of qualified persons prepared pursuant to the annexed Statute.

Article 6

The jurisdiction of the Tribunal shall comprise all disputes submitted to it in accordance with this Chapter and all matters specifically provided for in other Chapters of this Convention or in any other international agreement, public or private, which confers jurisdiction on the Tribunal with respect to a dispute relating to the interpretation or application of this Convention.

Article 7

The decisions of the Tribunal shall be binding upon the parties.

Article 8

1. The Tribunal shall expeditiously handle disputes which are of an urgent character and shall in appropriate cases issue binding interim orders for the purpose of minimizing damage to any party pending final adjudication. The Tribunal may also take such binding interim action in cases which have been submitted to arbitration under Articles 1 or 3.

2. The owner or operator of any vessel detained by any State shall have the right to bring the question of the detention of the vessel before the Tribunal in order to secure its prompt release in accordance with the applicable provisions of this Convention, without prejudice to the merits of any case against the vessel.

3. Nothing in these Articles shall abridge the sovereign immunity to which certain vessels and aircraft are entitled under international law.

Article 9

Any undertaking to have recourse to a particular means for compulsory settlement of a dispute relating to the interpretation or application of this Convention, whether contained in a general, regional or special international agreement, a license, lease or contract, constitutes an obligation to be carried out in good faith.

BIBLIOGRAPHY

BOOKS

Alexander, Lewis M. The Law of the Sea: Offshore Boundaries and Zones. Columbus: Ohio State University Press, 1967.

Allison, Graham. Essence of Decision: Explaining the Cuban Missile Crisis. Boston: Little, Brown and Company, 1971.

Auguste, Barry. The Continental Shelf: The Practice and Policy of the Latin American States With Special Reference to Chile, Ecuador, and Peru. Geneva: Librairie E. Droz, 1960.

Azcarraga, Jose Luis de. La Plataforma Submarina y el Derecho Internacional. Madrid: Universidad de Madrid, 1952.

Bartley, Ernest R. The Tidelands Oil Controversy. Austin: University of Texas Press, 1953.

Batelle Memorial Institute. Legal Aspects of the Continental Shelf and Seabed. ACDA/ST-120, Columbus, Ohio, 1967.

Braybrooke, David and Charles E. Lindblom. A Strategy of Decision. New York: The Free Press, 1970.

Brown, Edward Duncan. Arms Control in Hyrdospace: Legal Aspects. Washington, D.C.: Woodrow Wilson International Center for Scholars, June 1971.

Bureau of Intelligence and Research, U.S. Department of State. Sovereignty of the Sea. Geographic Bulletin no. 3, April 1965.

Burke, William T. International Legal Problems of Scientific Research in the Oceans. Springfield, Virginia: Clearinghouse for Federal Scientific and Technical Information, August 1967, PB 177 724.

________. Towards A Better Use of the Ocean. New York: Humanities Press, 1969.

Butler, William E. The Soviet Union and the Law of the Sea. Baltimore: The Johns Hopkins Press, 1971.

Carr, E. H. The Twenty Years Crisis, 1919-1939. New York: Harper and Row, 1964.

Colombos, C.J. The International Law of the Sea. 5th ed. London: Longman's Green, 1962.

Commission on Marine Science, Engineering and Resources. Industry and Technology. Washington, D.C.: G.P.O.; 1965.

________. Marine Resources and Legal-Political Arrangements for Their Development. Washington, D.C.: G.P.O., 1969.

________. Our Nation and the Sea: A Plan for National Action. Washington, D.C.: G.P.O., 1969.

________. Science and Environment. Washington, D.C.: G.P.O., 1969.

Crocker, Henry Graham. The Extent of the Marginal Sea. Washington, D.C.: G.P.O., 1919.

Deutsch, Karl and Stanley Hoffman. The Relevance of International Law. Garden City N.Y.: Doubleday, 1971.

Downs, Anthony. Inside Bureaucracy. Boston: Little, Brown and Company, 1967.

Engler, Robert. The Politics of Oil: A Study of Private Power and Democratic Directions. Chicago: University of Chicago Press, 1967.

Fenno, Jr., Richard F. The Power of the Purse: Appropriation Politics in Congress. Boston: Little, Brown and Company, 1966.

Friedmann, Wolfgang. The Changing Structure of International Law. New York: Columbia University Press, 1964.

________. The Future of the Oceans. New York: George Braziller, 1971.

Garcia Amador, F. W. The Exploitation and Conservation of the Resources of the Sea. Leyden, Netherlands: A. W. Sythoff, 1963.

Gidel, Gilbert. La Plataforma Continental Ante el Derecho. Universidad de Valladolid, 1951.

Hart, H. L. A. The Concept of Law. London: Oxford University Press, 1961.

Henkin, Louis. How Nations Behave. New York: Praeger Publishers, 1968.

________. Law for the Sea's Mineral Resources. Washington, D.C.: National Council on Marine Resources and Engineering Development, December 1967.

Hoffman, Stanley. The State of War. New York: Praeger Publishers, 1965.

Jane's Fighting Ships 1957-1958. London: Jane's Fighting Ships Publishing Company, 1958.

Jessup, P. C. The Law of Territorial Waters and Maritime Jurisdiction. New York: G. A. Jennings Company, 1927.

Johnston, Douglas M. The International Law of Fisheries. New Haven: Yale University Press, 1965.

Kaplan, Morton and Nicholas Katzenbach. The Political Foundations of International Law. New York: John Wiley, 1961.

Key Jr., V. O. Politics, Parties, and Pressure Groups. 5th ed. New York: Thomas Y. Crowell, 1964.

Lasswell, Harold. Politics: Who Gets What, When, How. New York: McGraw-Hill, 1936.

Leonard, A. L. International Regulation of Fisheries. Concord, New Hampshire: Rumford Press, 1944.

Lissitzyn, Oliver J. International Law Today and Tomorrow. Dobbs Ferry, New York: Oceana Publications, 1965.

Logue, John J., ed. The Fate of the Oceans. Villanova: Villanova University Press, 1972.

Lopez Villamil, Humberto. La Plataforma Continental y los Problemas Juridicos del Mar. Madrid, 1958.

Martin, L. W. The Sea in Modern Strategy. New York: Praeger Publishers for the Institute of Strategic Studies, 1967.

McDougal, Myres S. and William T. Burke. The Public Order of the Oceans. New Haven: Yale University Press, 1962.

Mero, John L. The Mineral Resources of the Sea. New York: Elsevier, 1965.

Mikdashi, Zuhayr. The Community of Oil Exporting Countries. Ithaca: Cornell University Press, 1972.

Mouton, M. W. The Continental Shelf. The Hague: Martinus Nijhoff, 1952.

National Academy of Sciences. Oceanography 1966: Achievements and Opportunities. A Report of the Committee on Oceanography Washington, D.C.: National Research Council, 1967.

National Petroleum Council. Law of the Sea: Particular Aspects Affecting the Petroleum Industry. Washington, D.C., May 1973.

________ Ocean Petroleum Resources: An Interim Report of the National Petroleum Council. Washington, D.C., July 4, 1974.

________ Petroleum Resources Under the Ocean Floor: An Interim Report of the National Petroleum Council. Washington, D.C., July 9, 1968.

________ Petroleum Resources Under the Ocean Floor. Washington, D.C., 1969.

________ Petroleum Resources Under the Ocean Floor: Supplemental Report of the National Petroleum Council. Washington, D.C., March 1971.

The Navy and the Common Sea. Washington, D.C.: Office of Naval Research, 1972.

Neustadt, Richard. Presidential Power. New York: Signet Books, 1960.

Oda, Shigeru. International Control of Sea Resources. Leyden, Netherlands: A. W. Sijthoff, 1963.

Oxman, Bernard. The Preparation of Article 1 of the Convention on the Continental Shelf. Washington, D.C.: U.S. Department of Commerce, National Bureau of Standards, prepared for the Commission on Marine Science, Engineering and Resources, 1968.

President's Science Advisory Committee. Effective Use of the Sea. Report of the Panel on Oceanography of the President's Science Advisory Committee. Washington, D.C.: G.P.O., June 1966.

Reiff, Henry. The United States and the Treaty Law of the Sea. Minneapolis: University of Minnesota Press, 1959.

Rosenau, James. Linkage Politics. New York: The Free Press, 1969.

________. National Leadership in Foreign Policy. Princeton: Princeton University Press, 1963.

________. The Scientific Study of Foreign Policy. New York: The Free Press, 1971.

Sapin, Burton. The Making of United States Foreign Policy. Washington, D.C.: The Brookings Institution, 1966.

Scelle, Georges. Plateau Continental et Droit International. Paris: A. Pedone, 1955.

Shalowtiz, Aaron. Shore and Sea Boundaries. Two volumes. Washington, D.C.: G.P.O., 1962.

Slouka, Zdenek J. International Custom and the Continental Shelf. The Hague: Martinus Nijhoff, 1968.

Smith, H. A. The Law and Custom of the Sea. London: Stevens, 1961.

Suarez, José Leon. El Mar Territorial y las Industrias Maritimas. Buenos Aires, 1918.

Sztucki, Jerezy, ed. Symposium on the International Regime of the Sea-Bed, Rome, June 30-July 5, 1969. Rome: Academia Nazionale Dei Lincei, 1970.

Tellez Benoit, Maria Emilia. La Plataforma Continental. Mexico City: Universidad Nacional Autonoma de Mexico, 1946.

Truman, David B. The Governmental Process. New York: Alfred A. Knopf, Inc., 1951.

Wenk, Jr., Edward. The Politics of the Ocean. Seattle: University of Washington Press, 1972.

Wildavsky, Aaron. The Politics of the Budgetary Process. Boston: Little, Brown and Company, 1964.

Wolfke, Karol. Custom in Present International Law. Wroclaw: Zahlad Narodowy im Ossolifiskich, 1964.

Woll, Peter. American Bureaucracy. New York: W.W. Norton, 1963.

ARTICLES

Adams, Matthew T. "U.S. Tax Aspects to Seabed Operations." Law and Policy in International Business 2 (Summer 1970): 443-478.

Allison, Graham and Morton Halperin. "Bureaucratic Politics: A Paradigm and Some Policy Implications." World Politics 24 (Spring 1972): 40-79.

Andrassy, Juraj. "Present Regime of the Military Uses of the Sea-Bed; Possible Regimes to be Envisaged." In Sztucki, J. ed. Symposium on the International Regime of the Sea-Bed. (Rome: Academia Nazionale Dei Lincei, 1970).

Aramburu y Monchaca, Andres A. "Character and Scope of the Rights Declared and Practiced Over the Continental Sea and Shelf." American Journal of International Law 47 (1953): 120-123.

Armstrong, P. B. "Anti-Submarine Warfare." Oceanology International, June 15, 1969, pp. 22-23.

Askevold, Gerald. "Ocean Mining in Perspective." Stanford Journal of International Studies 4 (1969): 115-142.

Auerbach, Carl A. "Comparison Between Recommendations of President Nixon and Those of the Commission on Marine Science,

Engineering and Resources on the Mineral Resources Underlying the High Seas." Proceedings of the Fifth Annual Conference of the Law of the Sea Institute, June 15-19, 1970. University of Rhode Island, 1971, p. 183.

________. "Tentative Recommendations of the Commission on Marine Science, Engineering and Resources." Proceedings of the Third Annual Law of the Sea Institute, June 24-27, 1968. University of Rhode Island, 1969, pp. 4-11.

Azzam, Issam. "Dispute Between France and Brazil Over Lobster Fishing in the Atlantic." International and Comparative Law Quarterly 13 (1964): 1453-1459.

Bailey, Kenneth. "Australia and the Law of the Sea." Adelaide Law Review 1 (1966): 1-22.

Becker, Loftus. "Just Compensation in Expropriation Cases: Decline and Partial Recovery." U.S. Department of State Bulletin 40 (June 1, 1959): 784-792.

Belman, Murray J. "The Role of the State Department in Formulating Federal Policy Regarding Marine Resources." Natural Resources Lawyer 1 (June 1968): 14-22.

Bennett, Elmer. "Legal Climate for Underseas Mining." Transactions, Marine Technology Society (1966): 204-210.

Bernfeld, Seymour. "Developing the Resources of the Sea—Security of Investment." International Lawyer 2 (1967): 67-76.

________. "The Mining Industry and the Continental Shelf Convention." Institution of Mining and Metallurgy 78 (1969): A10-A19.

Bingham, Joseph W. "Changing Concepts of International Law: Maritime Jurisdiction in Time of Peace." Proceedings of the American Society of International Law 34 (1940): 54-63.

________. "The Continental Shelf and the Marginal Belt." American Journal of International Law 40 (1946): 173-178.

________. "Pacific Coastal Fisheries." American Bar Association. Section on International and Comparative Law, Selected Papers and Reports (1940): 46-48.

Bishop Jr., William W. "The Need for a Japanese Fisheries Agreement." American Journal of International Law 45 (1951): 712-719.

Blanchard, Fletcher and Robert W. Corell. "Man's Adaptation to the Sea: A Key to the Ocean's Resources." Stanford Journal of International Studies 4 (1969): 71-83.

Boggs, S. W. "Delimitation of Seaward Areas Under National Jurisdiction." American Journal of International Law 45 (1951): 240-266.

Borchard, Edwin. "Jurisdiction Over the Littoral Bed of the Sea." American Journal of International Law 35 (1941): 515-519.

________. "Resources of the Continental Shelf." American Journal of International Law 40 (1946): 53-70.

Borgese, Elisabeth Mann. "The Ocean Regime." A Center Occasional Paper, vol. 1, no. 5 (Santa Barbara: Fund for the Republic, October 1968).

Breckner, Norman V. "Some Dimensions of Defense Interest in the Legal Delimitations of the Continental Shelf." Proceedings of the Fourth Annual Conference of the Law of the Sea Institute. Kingston: University of Rhode Island, 1970, pp. 188-192.

Briggs, H. W. "Jurisdiction Over the Seabed and Subsoil Beyond Territorial Waters." American Journal of International Law, 39 (1945): 338-342.

Brock, John. "Mineral Resources and the Future Development of the International Law of the Sea." JAG Journal 22 (September-October-November 1967): 39-44.

Brooke, David B. "Deep Sea Manganese Nodules: From Scientific Phenomenon to World Resources." Proceedings of the Second Annual Conference of the Law of the Sea Institute. Kingston: University of Rhode Island, 1968, pp. 32-41.

Brown, E. D. "The Legal Regime of Inner Space—Military Aspects." Current Legal Problems 22 (1969): 181-204.

Brownlie, Ian. "Recommendations on the Limits of the Continental Shelf and Related Matters: A Commentary." Proceedings of the Fourth Annual Conference of the Law of the Sea Institute. Kingston: University of Rhode Island, 1970, p. 133.

Bullington, John, Edward Allen and Richard Burges. "Title to Petroleum Beneath the Bed of the Sea." American Bar Association, Section on International and Comparative Law, Selected Papers and Reports (1940), pp. 44-46.

Burke, William T. "Legal Aspects of Ocean Exploitation—Status and Outlook." Transactions of the Second Annual Marine Technology Society Conference, June 27-29, 1966, pp. 1-23.

________. "A Negative View of a Proposal for United Nations Ownership of Ocean Mineral Resources.": Natural Resources Lawyer 1 (June 1968): 42-62.

________. "Ocean Sciences and the Future International Law of the Sea." Morshon National Security Program: Ohio State University Press, 1966.

Campbell, Nicholas J. "International Law Developments Concerning National Claims to and in Offshore Areas." Tulane Law Review 33 (1958-1959): 339-356.

Carter, Jared. "The Seabed Beyond the Limits of National Jurisdiction." Stanford Journal of International Studies 4 (1969): 1-31.

Castenada, Jorge. "Underdeveloped Nations and the Development of International Law." International Organization 15 (1961): 38-48.

Chapman, Wilbert. "Conflict of Uses of the Sea." Proceedings of the Second Annual Conference of the Law of the Sea Institute. Kingston: University of Rhode Island, 1968, pp. 70-71.

________. "The Ocean Regime of the Real World." Proceedings of the Fourth Annual Conference of the Law of the Sea Institute. Kingston: University of Rhode Island, 1970, pp. 446-469.

________. "The United States Fish Industry and the 1958 and 1960 United States Conference on the Law of the Sea." Proceedings of the Third Annual Conference of the Law of the Sea Institute. Kingston: University of Rhode Island, 1969, pp. 35-63.

________. "United States Policy on High Seas Fisheries." U.S. Department of State Bulletin 20 (January 16, 1949): 67-71, 80.

Cheever, Daniel S. "Marine Science and Ocean Politics." Science and Public Affairs (February 1970), p. 22.

Christopher, W. M. "The Outer Continental Shelf Lands Act: Key to a New Frontier." Stanford Law Review 6 (1953-1954): 23-68.

Christy Jr., Francis T. "Fisheries and the New Conventions on the Law of the Sea." San Diego Law Review 7 (1970): 455-468.

Clark, Harold F. and George T. Renner. 'We Should Annex 50,000,000 Square Miles of Ocean." Saturday Evening Post May 4, 1946, pp. 16.

Cohen, Paul. "The Erosion of Surface Naval Power." Foreign Affairs 49 (January 1971): 330-341.

________. "The Future of the Submarine." Foreign Affairs 38 (October 1959): 110-120.

Cosford, Edwin J. "The Continental Shelf 1910-1945." McGill Law Journal 4 (Spring 1958): 245-266.

Craven, John. "Sea Power and the Sea Bed." U.S. Naval Institute Proceedings 92 (April 1966): 36-51.

________. "Technology and the Law of the Sea." Ohio State University Conference on Law, Organization and Security in the Use of the Ocean 2 (March 17-18, 1967): 35-36.

Crawford, John E. and John W. Pedan. "The Bureau of Mines' Expanding Role in Undersea Mining." Mining Engineering (March 1965): 67.

Dean, Arthur H. "Geneva Conference on the Continental Shelf." Tulane Law Review 41 (1966-67): 419-432.

________. "Geneva Conference on the Law of the Sea." American Journal of International Law 52 (1958): 607-628.

Dole, Hollis. "Ocean Minerals and the Law." Natural Resources Lawyer 2 (November 1969): 352-359.

Ely, Northcutt. "American Policy Options in the Development of Undersea Mineral Resources." International Lawyer 2 (1967-68): 215-223.

________. "A Case for the Administration of Mineral Resources Underlying the High Seas by National Interests." Natural Resources Lawyer 1 (June 1968): 78-84.

Falk, Richard. "The Adequacy of Contemporary Theories in International Law-Gaps in Legal Thinking." Virginia Law Review, 50 (1964): 231-265.

Fatouros, A. A. "International Law and the Third World." Virginia Law Review 50 (1964): 783-823.

Feller, Peter B. "U.S. Customs Aspects of Seabed Operations." Law and Policy in International Business 2 (Summer 1970): 402-442.

Finlay, Luke. "Critique of the Marine Science Commission Recommendations." Proceedings of the Fourth Annual Conference of the Law of the Sea Institute. Kingston: University of Rhode Island, 1970, pp. 50-55.

Francois, J. P. A. "Some Aspects of the Extension of National Sovereignty of the Adjacent Sea." International Relations (April 1955): 79-83.

Friedheim, R. L. "The Continental Shelf Issue at the United Nations: A Quantitative Content Analysis." Professional Paper No. 7. Center for Naval Analyses (January 7, 1970).

________. "Understanding the Debate on Ocean Resources." Law of the Sea Institute. University of Rhode Island, Occasional Paper no. 1 (February 1968).

Friedmann, Wolfgang. "The Race to the Bottom of the Sea." Columbia Forum (Spring 1968): 18-21.

Frosch, Robert A. "Exploiting Marine Mineral Resources—Problems of National Security and Jurisdiction." Defense Industry Bulletin 4 (September 1968): 10.

________. "Marine Mineral Resources: National Security and National Jurisdiction." Naval War College Review 21 (1968): 53-60.

________. "The Military Uses of the Ocean." Second Conference on Law, Organization and Security in the Use of the Ocean. October 5-7, 1967. Ohio State University, 1967.

Garrett, Michael L. "Issues in International Law Created by Scientific Development of the Ocean Floor." Southwestern Law Journal 19 (March 1965): 97-115.

George, Alexander L. "The Case for Multiple Advocacy in Making Foreign Policy." American Political Science Review (September 1972): 751-785.

Goddard, G. "The Land Under the Sea." Indian Yearbook of International Affairs 6 (1957): 81-103.

Gottleib, A. E. "Recent Developments Concerning Exploration and Exploitation of the Ocean Floor." McGill Law Journal 15 (1969): 260-278.

Green, L. C. "The Geneva Conventions and the Freedom of the Seas." Current Legal Problems 12 (1959): 224-246.

Griffin, William L. "Development of Law for Ocean Activities." Transactions, Marine Technology Society (1966): 348.

Griffin, William L. "The Emerging Law of Ocean Space." International Lawyer 1 (1967): 548-587.

Grunwalt, Richard J. "The Acquisition of the Resources of the Bottom of the Sea, A New Frontier of International Law." Military Law Review (October 1966): 101-133.

Gutteridge, J. A. C. "The 1958 Convention on the Continental Shelf." British Yearbook of International Law 35 (1959): 102-123.

Hagen, Virginia M. "The Latin American—United States Fishing Rights Controversy: Dilemma for United States Foreign Policy (1969-1970)." Library of Congress, Congressional Research Service (April 15, 1971).

Hearn, Wilfred. "The Fourth Dimension of Seapower—Ocean Technology and International Law." JAG Journal 22 (September-October-November 1967): 23-25.

________. "The Role of the United States Navy in the Formulation of Federal Policy Regarding the Sea." *Natural Resources Lawyer* 1 (1968): 23-31.

Hedberg, Hollis D. "Limits of National Jurisdiction Over Natural Resources of the Ocean Bottom." *Proceedings of the Fourth Annual Conference of the Law of the Sea Institute*. Kingston: University of Rhode Island, 1970, pp. 159-169.

Henkin, Louis. "International Law and 'the Interests': The Law of the Seabed." *American Journal of International Law* 63 (1969): 504-509.

Herrington, William C. "The Convention on Fisheries and Conservation of Living Resources: Accomplishments of the 1958 Geneva Conference." In Alexander, L. M. *The Law of the Sea: Offshore Boundaries and Zones*. Ohio: Ohio State University, Press 1967.

________. "U.S. Policy on Fisheries and Territorial Waters." *U.S. Department of State* Bulletin 26 (June 30, 1952): 1021-1023.

Hersh, Seymour. "An Arms Race on the Sea Bed?" *War/Peace Report* (August-September 1968).

Hibbard Jr., Walter L. "The Government's Program for Encouraging the Development of a Marine Mining Industry." *Transactions*, Marine Technology Society (1966): 197-203.

Hilsman, Roger. "The Foreign-Policy Consensus: An Interim Research Report." *Journal of Conflict Resolution* 3 (1959): 361.

Hirdman, Sven. "Weapons in the Deep Sea." *Environment* 13 (April 1971): 28.

Holden Jr., Matthew. " 'Imperialism' in Bureaucracy." *American Political Science Review* 60 (December 1966): 943-951.

Holland, Henry F. "The Juridical Status of the Continental Shelf." *Texas Law Review* 30 (1951-1952): 586-598.

Hollick, Ann. "Seabeds Make Strange Politics." *Foreign Policy* 9 (1972-1973): 148-170.

________. "United States Oceans Politics." *San Diego Law Review* 10 (1973): 467-501.

Hounshell, C. D. and L. H. Kemp. "The Continental Shelf: A Study in National Interests and International Law." *Journal of Public Law* (Fall 1956): 343-362.

"How Do We Get the Independent Offshore?" *Petroleum Independent* (November-December 1971): 16-21.

Hull, E. W. S. "The Political Ocean." *Foreign Affairs* 45 (April 1967): 492-502.

Hurst, C. "Whose is the Bed of the Sea?" *British Yearbook of International Law* (1923-1924): 34-43.

Ickes, Harold. "Underwater Wealth." *Collier's*, February 23, 1946, p. 20.

Idyll, Clarence P. "The Crab That Shakes Hands." *National Geographic* 139 (February 1971): 254-271.

Jennings, R. Y. "The Limits of Continental Shelf Jurisdiction: Some Possible Implications of the North Sea Case Judgment." *International and Comparative Law Quarterly* 18 (1969): 819-832.

Kahn Jr., E. J. "The Catch in the Hatch Isn't What it Used to Be." *New York Times Magazine*, August 29, 1971.

Kassel, B. M. "The Fishing Fleet of the Soviet Union." *United States Naval Institute Proceedings* 87 (1961): 70-81.

Knauss, John A. "The Military Role in the Ocean and its Relations to the Law of the Sea." (June 1, 1971).

Knoll, Erwin. "The Oil Lobby is Not Depleted." *New York Times Magazine*, March 8, 1970.

Koh, Kwang Lim. "The Continental Shelf and the International Law Commission." *Boston University Law Review* 35 (1955): 522-540.

Krasner, Stephen D. "Are Bureaucracies Important?" *Foreign Policy* (Summer 1972): 159-179.

Krueger, R. "State Tidelands Leasing in California." *U.C.L.A. Law Review* 5 (1958): 427-489.

Kunz, J. L. "The Continental Shelf and International Law: Confusion and Abuse." American Journal of International Law 50 (1956): 828-853.

Lauterpacht, H. "Sovereignty over Submarine Areas." British Yearbook of International Law 27 (1950): 376-433.

Lawrence, Keith. "Military-Legal Considerations in the Extension of the Territorial Sea." Military Law Review 29 (1965): 47-95.

Lawson, Carol. "Building a Regime for the Sea: Moving Into Deep Water." War/Peace Report (November 1971): 10-11.

Levy, Walter J. "Oil Power." Foreign Affairs 49 (July 1971): 652-668.

Luard, Evan. "Who Gets What on the Seabed." Foreign Policy 9 (1972-1973): 132-147.

Luce, Charles F. "The Development of Ocean Minerals and the Law of the Sea." Natural Resources Lawyer 1 (July 1968): 29-35.

MacGibbon, I. C. "The Scope of Acquiescence in International Law." British Yearbook of International Law 31 (1954): 143-186.

McDevitt, Joseph. "New Issues and New Interests in the Law of the Sea." Naval War College Review 21 (1969): 12-29.

Mero, John L. "Alternatives for Mineral Exploitation." Proceedings of the Second Annual Conference of the Law of the Sea Institute. Kingston: University of Rhode Island, 1968, pp. 94-97.

________. "Mineral Deposits in the Sea." Natural Resources Lawyer 1 (July 1968): 130-137.

Miron, George. "The Management of the Mineral Resources of the Ocean Floor—A Critique of Certain Aspects of the Proposal of the Commission on Marine Science, Engineering and Resources." Stanford Journal of International Studies 4 (1969): 32-45.

Nanda, V. P. and K. R. Stiles. "Offshore Oil Spills: An Evaluation of Recent United States Responses." San Diego Law Review 7 (1970): 519-540.

Nicholson, William M. "A Navy View of Ocean Resources." Natural Resources Lawyer 1 (January 1968): 77-81.

Nierenberg, William A. "Military Oceans." In Calder, Nigel, Unless Peace Comes. London: Penguin, 1968, pp. 111-119.

O'Connor, Dennis M. "Critique of the Marine Science Commission Recommendations." Proceedings of the Fourth Annual Law of the Sea Institute. Kingston: University of Rhode Island, 1970.

Oda, S. "The Continental Shelf." The Japanese Annual of International Law 15, (1957): 15-37.

________. "A Reconsideration of the Continental Shelf Doctrine." Tulane Law Review (December 1957): 21-36.

Oswald, James. "Toward a Political Theory of the Ocean." Transactions, Marine Technology Society (1966).

Oxman, Bernard H. "Report on the President's Proposal." Proceedings of the Fifth Annual Conference of the Law of the Sea Institute. Kingston: University of Rhode Island, 1971, p. 160.

Padwa, D. J. "Submarine Boundaries." The International and Comparative Law Quarterly 9 (October 1960): 628-653.

Palmer, William. "The United States Draft United Nations Convention on the International Seabed Area and the Accommodation of Ocean Uses." Syracuse Journal of International Law and Commerce 1 (1972): 110-115.

Pardo, Arvid. "Who Would Control the Seabed." Foreign Affairs 47 (October 1968): 122-137.

Pehrson. "Mining Industry's Role in Development of Undersea Mining." Transactions, Marine Technology Society (1966), p. 182.

Phister, Montgomery. "Regime of the High Seas." Proceedings of the American Society of International Law 50 (1956): 136-154.

Phleger, H. "Recent Developments Affecting the Regime of the High Seas." Department of State Bulletin 32 (June 6, 1955): 934-940.

Powers, Robert D. and Leonard Hardy. "How Wide the Territorial Sea?" In United States Naval War College International Law Studies, 1959-1960. Washington, D.C.: G.P.O., 1961, pp. 304-312.

Purrett, Louise. "The Politics of Marine Research." Science News 99 (January 2, 1971): 9-11.

Ratiner, Leigh. "United States Oceans Policy: An Analysis." Journal of Maritime Law and Commerce 2 (January 1971): 225-266.

Ratiner, Leigh S. and L. Rebecca Wright. "United States Ocean Mineral Resource Interests and the United Nations Conference on the Law of the Sea." Natural Resources Lawyer 6 (1973): 1-42.

Robertson, J. W. "Security Interests and Regimes of the Sea." in Alexander, L. M., ed. The Law of the Sea: The Future of the Sea's Resources. Kingston: University of Rhode Island, 1968, pp. 47-51.

Schaefer, Milner B. "The Changing Law of the Sea—Effects on Freedom of Scientific Investigation." Proceedings of the Second Annual Conference of the Law of the Sea Institute (June 26-29, 1967): 113-120.

________. "Freedom of Scientific Research and Exploration in the Sea." Stanford Journal of International Studies 4 (1969): 46-70.

________. "Some Recent Developments Concerning Fishing and the Conservation of the Living Resources of the High Sea." San Diego Law Review 7 (1970): 371-407.

"Sea Mounds May Lead to Deep Water Oil Resources." Oil and Gas Journal, November 26, 1956.

Slouka, Zdenek. "United Nations and the Deep Ocean: From Date to Norms." Syracuse Journal of International Law and Commerce 1 (1972): 61-90.

Smith, Benjamin. "Our Fishing Fleet Comes in Fifth." New York Times Magazine, October 7, 1962.

Stevens, Theodore F. "A United States Policy for the Wet Frontier." Natural Resources Lawyer (November 1969): 347-351.

Stevenson, John R. and Bernard Oxman. "The Preparations for the Law of the Sea Conference." American Journal of International Law 68 (1974): 1-32.

Stone, Oliver L. "The Marine Environment—Recent Legal Developments." Natural Resources Lawyer 2 (January 1969): 26-46.

Stone, Oliver L. "United States Legislation Regulating the Continental Shelf." International and Comparative Law Quarterly 17 (1968): 103-117.

Sullivan Jr., William L. "Freedom of Scientific Research in Conflict with the Development of Ocean Resources." Presented at the American Society of International Law, Regional Meeting, Miami, December 13, 1969.

Swan, Charles and James Ueberhorst. "The Conference on the Law of the Sea." Michigan Law Review 56 (1957-1958): 1132-1141.

Tubman, William C. "The Legal Status of Minerals Located on or Beneath the Ocean Floor Beyond the Continental Shelf." Transactions, Marine Technology Society (1966), pp. 379-404.

Vallat, F. A. "The Continental Shelf." British Yearbook of International Law 23 (1946): 333-338.

Waldock, C. H. M. "The Legal Basis of Claims to the Continental Shelf." Transactions of the Grotius Society 36 (1950): 115-148.

Wedin, John. "Impact of Distant Water Fishermen on Coastal Fisheries." Proceedings of the Second Annual Conference of the Law of the Sea Institute. Kingston: University of Rhode Island, 1968, pp. 14-18.

Wenk Jr., Edward. "Congress Sharpens Ocean Interests." Undersea Technology (January 1966): 36-42.

Whiteman, Marjorie. "Conference on the Law of the Sea: Convention on the Continental Shelf." American Journal of International Law 52 (1958): 629-659.

________. "The Territorial Sea." Proceedings of the American Society of International Law 50 (1956): 125-136.

Windley, David W. "International Practice Regarding Traditional Privileges of Foreign Fishermen in Zones of Extended Maritime Jurisdiction." American Journal of International Law 63 (1969): 409-503.

Young, Richard. "The Continental Shelf in the Practice of American States, 1950-1951." Inter-American Juridical Yearbook (1953): 27-36.

________. "The Geneva Convention on the Continental Shelf: A First Impression." American Journal of International Law 52 (1958): 733-738.

________. "The Legal Status of Submarine Areas." American Journal of International Law 45 (1951): 225-239.

________ "The Over-Extension of the Continental Shelf." American Journal of International Law 47 (1953): 454-456.

________. "Pan American Discussion of Offshore Claims." American Journal of International Law 50 (1956): 909-916.

________. "Recent Developments with Reference to the Continental Shelf." American Journal of International Law 42 (1948): 849-857.

CONGRESSIONAL HEARINGS AND REPORTS

Hearings. House Committee on Armed Services, Subcommittee on Seapower. "Territorial Sea Boundaries." 91st Congress, 2d session, 1970.

Hearings. House Committee on Foreign Affairs, Subcommittee on International Organizations and Movements. "Law of the Sea and Peaceful Uses of the Seabed." 92nd Congress, 2d session, 1972.

Hearings. House Committee on Foreign Affairs, Subcommittee on International Organizations and Movements. "Law of the Sea Resolution." 93rd Congress, 1st session, 1973.

Hearings. House Committee on the Judiciary. "Submerged Oil Lands." 76th Congress, 1st session, 1939.

Hearings. House Committee on the Judiciary. "Title to Submerged Oil Lands." 75th Congress, 3rd session, 1938.

Hearings. House Committee on Merchant Marine and Fisheries. "Alaska Salmon Fishery." 75th Congress, 3rd session, 1938.

Hearings. House Committee on Merchant Marine and Fisheries. "Fishing in United States Territorial Waters." 88th Congress, 2d session, 1964.

Hearings. House Committee on Merchant Marine and Fisheries. "National Oceanographic Program Legislation." 89th Congress, 1st session, 1965.

Hearings. House Committee on Merchant Marine and Fisheries, Subcommittee on Fisheries and Wildlife Conservation. "Miscellaneous Fisheries Legislation Fishing Rights." 89th Congress, 2d session, 1966.

Hearings. House Committee on Merchant Marine and Fisheries, Subcommittee on Fisheries and Wildlife Conservation and the Environment. "Territorial Seas." 93rd Congress, 1st session, 1973.

Hearings. House Committee on Merchant Marine and Fisheries, Subcommittee on Oceanography. "Deep Seabed Hard Mineral Resources." 92nd Congress, 2d session, 1972.

Hearings. House Committee on Merchant Marine and Fisheries, Subcommittee on Oceanography. "Deep Seabed Hard Minerals." 93rd Congress, 1st and 2nd sessions, 1973.

Hearings. House Committee on Merchant Marine and Fisheries, Subcommittee on Oceanography. "Oceanography Miscellaneous." 92nd Congress, 2d session, 1972.

Hearings. Senate Committee on Commerce, Subcommittee on Merchant Marine and Fisheries. "Fishing in United States Territorial Waters." 88th Congress, 1st session, 1963.

Hearings. Senate Committee on Commerce. Subcommittee on Oceans and Atmosphere. "Interim Fisheries Zone Extension and Management Act of 1973." 93rd Congress, 1st and 2d sessions 1973-74.

Hearings. Senate Committee on Commerce. "Special Study on United Nations Suboceanic Lands Policy." 91st Congress, 1st session, 1969.

Hearings. Senate Committee on Foreign Relations, Subcommittee on Ocean Space. "Activities of Nations in Ocean Space." 91st Congress, 1st session, 1969.

Hearings. Senate Committee on Foreign Relations. "Conventions on the Law of the Sea." 86th Congress, 2d session, 1960.

Hearings. Senate Committee on Foreign Relations. "Governing the Use of Ocean Space." 90th Congress, 1st session, 1967.

Hearings. Senate Committee on Interior and Insular Affairs. "Outer Continental Shelf." 83rd Congress, 1st session, 1953.

Hearings. Senate Committee on Foreign Relations. "Seabed Arms Control Treaty." 92nd Congress 2d, session, 1972.

Hearings. Senate Committee on Foreign Relations, Subcommittee on Oceans and International Environment. "U.S. Oceans Policy." 93rd Congress, 1st session, 1973.

Hearings. Senate Committee on Interior and Insular Affairs, Subcommittee on Minerals, Materials and Fuels. "Mineral Resources of the Deep Seabed." 93rd Congress, 1st session, 1973.

Hearings. Senate Committee on Interior and Insular Affairs. "Submerged Lands." 83rd Congress, 1st session, 1953.

Hearings. Senate Committee on Interior and Insular Affairs, Special Subcommittee on the Outer Continental Shelf. "Outer Continental Shelf." 91st Congress, 1st and 2d sessions, 1969-1970.

Hearings. Senate Committee on Interior and Insular Affairs, Subcommittee on Minerals, Materials and Fuels. "Status Report on Law of the Sea Conference." 93rd Congress, 1st session, 1973.

Hearings. Senate Committee on Public Lands and Surveys. "Title to Submerged Lands." 76th Congress, 1st session, 1939.

House Report. Committee on Foreign Affairs, Subcommittee on International Organization and Movements. House Report no. 1957. "The Oceans: A Challenging New Frontier." 90th Congress, 2d session, October 9, 1968.

House Report. Committee on Foreign Affairs. House Report no. 999. "The United Nations and the Issue of Deep Ocean Resources." 90th Congress, 1st session, 1967.

House Report. Committee on Interior and Insular Affairs. House Report no. 2515. "Report of the Committee on Interior Affairs Pursuant to House Resolution 676, 82nd Congress, Authorizing an Investigation and Study of the Seaward Boundaries of the United States." 82nd Congress, 2d session, January 1953.

House Report. Committee on the Judiciary. House Report no. 215. "Report to Accompany H.R. 4198, Submerged Lands Act." 83rd Congress, 1st session, 1953.

House Report. Committee on Merchant Marine and Fisheries. House Report no. 688. "Declaring the Lobster a Creature of the Continental Shelf." 93rd Congress, 1st session, 1973.

House Report. Committee on Merchant Marine and Fisheries. House Report no. 687. "Offshore Shrimp Fisheries Act of 1973." 93rd Congress, 1st session, 1973.

House Report. Committee on Science and Astronautics. House Report 2078. "Ocean Sciences and National Security." 86th Congress, 2d session, 1960.

Senate Report. Committee on Commerce. Senate Report no. 426. "Advancement of Marine Sciences—Marine Sciences and Research Act of 1961." 87th Congress, 1st session, 1961.

Senate Report. Committee on Commerce. Senate Report no. 528. "Marine Resources and Engineering Development Act of 1965." 89th Congress, 1st session, July 29, 1965.

Senate Report. Committee on Commerce. Senate Report no. 633. "Offshore Shrimp Fisheries Act of 1973." 93rd Congress, 1st session, 1973.

Senate Report. Committee on Commerce. Senate Report no. 1280. "To Accompany S.2218, a Bill to Establish a 12 Mile Fishery

Zone Off the United States Coast." 89th Congress, 2d session, 1966.

Senate Report. Committee on Interior and Insular Affairs. Senate Report no. 411. "Outer Continental Shelf Lands Act." 83rd Congress, 1st session, 1953.

Senate Report. Committee on Interior and Insular Affairs, Special Subcommittee on Outer Continental Shelf. "Report on Outer Continental Shelf." 91st Congress, 2d session, 1970.

INDEX

ABOUT THE AUTHOR

LAWRENCE JUDA is Assistant Professor of Political Science at Muhlenberg College in Allentown, Pennsylvania. Previous to his position at Muhlenberg College, Dr. Juda taught at Bronx Community College in New York City.

Dr. Juda received his B.A. degree, with honors in political science, from the City College of the City University of New York and his Ph.D. in political science from Columbia University.

RELATED TITLES
Published by
Praeger Special Studies

INTERNATIONAL ENVIRONMENTAL LAW
edited by Ludwik A. Teclaff
and Albert Utton

THE INTERNATIONAL POLITICS OF MARINE
POLLUTION CONTROL
Robert A. Shinn

SINO-SOVIET TERRITORIAL DISPUTE
George Ginsburgs

TERRITORIAL SEAS AND INTER-AMERICAN
RELATIONS: With Case Studies of the Peruvian
and U.S. Fishing Industries
Bobbie B. and Robert M. Smetherman